The Economic Plants of North East India

Sudhir Kumar

SCIENTIFIC®
PUBLISHERS

Published by:

SCIENTIFIC PUBLISHERS (INDIA)
5A, New Pali Road, P.O. Box 91
Jodhpur 342 001, India
E-mail: info@scientificpubonline.com
Website: www.scientificpubonline.com

Print, 2022

ISBN: 978-81-72332-20-4
E-ISBN: 978-93-87893-36-8

Printed in India

THIS BOOK

IS

DEDICATED TO THE MEMORY OF

MY UNCLE

LATE SHRI UMA SHANKAR

WHO MADE ME

WHAT I AM

TODAY

FOREWORD

India is one of the largest mega centres of genetic diversity of plants. The Northeastern region of the country is, in particular, blessed with very rich biodiversity of both fauna and flora. Rapid developmental activity and fast changes of societal structure of this region are bound to have an impact on the continued sustenance of this diversity. There are many reports, some clearly alarming, that both species and genetic diversity in the area is under threat. The need for conservation of this rich national heritage is unanimous. However, if conservation is to be done on scientific basis, a data base of what actually exists today is of paramount importance. The efforts of Prof. Sudhir Kumar and his colleagues have tackled this particular aspect of the conservation requirement. 462 plants of economic importance have been studied and described. Concrete suggestions have been advanced for both conservation and exploitation of these plants of economic value. We owe a debt of gratitude for the effort that has gone into the study. The resulting compendium is a rich source of information. It is my hope that a large and diverse community of scientific interests will find the material in this publication both interesting and useful.

V.L. Chopra
Ex. Director General, I.C. A.R. New Delhi

PREAMBLE

The North Eastern Region of India is the fountain of rich diversity of Plants of Economic importance and richest reservoir of genetic variability. About 50% of the total number of species of our sub-continent occur in this region. The region is rich in endemic species and forms a "GATEWAY" for numerous alien species. Further the presence of a large number of primitive flowering plants makes N.E. India the "cradle of flowering plants" (Rao, 1993).

However, the area is still unexplored and information is rudimentary, infantile and scanty. It is high time that all scientific efforts are diverted towards identification, conservation, cultivation and in saving of these most valuable plant resources. These plants if properly exploited can change the entire economy of the region and result in establishment of a number of agroforestry based industries resulting in improving the quality of life of the people.

This compendium is a small effort to bring to the forth the plant wealth of the region. This compilation is particularly important in view of BIODIVERSITY CONVENTION as it lists the plant wealth of an important region of our country.

The author is fully conscious of the short comings of this compilation but will feel fully satisfied and amply rewarded if it stimulates further research on the subject and results in establishment of industries and thus removing regional imbalance.

2nd October, 1999

Sudhir Kumar
Farm House,
Bari/Sitapur - 261 303 India

CONTENTS

ACKNOWLEDGEMENT

The author expresses his grateful thanks to the **Department of Science and Technology, Government of India, New Delhi** and also to Dr. Parveen Farooqi, Head, Human Resource and Userf section for financial support for the preparation of this book. Author is grateful to Professor N.K. Mehrotra, Ex. Head of the Department of Botany, University of Lucknow for placing at his disposal the facilities of the department and the support he gave for this publication. He expresses his appreciation to Shri Sanjay Singh for help in collecting the data and Mrs. Geeta Mishra and Shri A.K. Singh for typing the manuscript.

The author is also grateful to the publisher Shri Pawan Kumar Sharma of Scientific Publishers (India), Jodhpur for immense help in publishing this book.

1

INTRODUCTION

The North Eastern Region of India comprising of Assam, Arunachal Pradesh, Manipur, Meghalaya, Mizoram, Nagaland and Tripura is situated between East longitude 89°42′ to 97°40′ and North latitude between 22°19′ and 29°50′. The rolling mountains, enchanting valleys, swift flowing streams and luxuriant forests because of plentiful rain throughout the year give evidence of precious gift of bountiful nature bestowed upon the region. Because of the wide geographical and climatic diversity, the region is also the repository of rich diversity of plants of economic importance and reservoir of genetic variability. This important plant wealth has neither been fully explored nor utilized to its full extent.

About 50% of the total number of species of our sub - continent occur in this region. The region is rich in endemic species and forms a "Gateway" to numerous alien species. Further the presence of large number of primitive flowering species makes the North East the cradle of flowering plants (Rao 1993).

However, this area is still unexplored and information on the plant resources available for this region is rather scanty. It is high time that all scientific efforts are directed towards identification, conservation and utilization of these valuable source of plant wealth which can provide the base line data for development of large number of plant based cottage industries. This will bring benefit to the area by improving economy of the region as well as contributing to the gross national product of the country. It will thus bring the benefit to the entire country. Many of the products developed from these plants could be import substitutes as well as important export items and thus saving millions of rupees of the country in foreign exchange.

The North Eastern Region has remained much neglected and there are thousands of promising species that await research and study, so that these plants can be utilized for the welfare of the people of India as well as the mankind. The cataloguing and detailed information with respect to utilization of their product will bring immense benefit to people of the region. This will also help in tapping new resources of fibres, natural dyestuff, fatty acids and essential oils etc. The natural dyestuff which had been abandoned due to use of cheaper synthetic dyestuff are coming in vogue since the European communities have banned the use of synthetic dyestuff. This has resulted in great loss to garment exporters of the country. The European economic community demands that the garments exported to these countries be dyed only in natural dyestuff, we therefore have to find out the new sources for the purpose. The informations to be obtained from tribals of the North Eastern Region will be of much help and value as they have been using natural dyestuff since time immemorial.

India has been importing large number of perfumes from Europe and other countries, resulting in depletion of foreign exchange. A number of essential oils found in North Eastern Region can be used as base for preparation of perfumes, thus saving valuable foreign exchange, and in the years to come we may be in a position to export new perfumes.

A number of food plants viz. vegetables and fruits are available only in the North Eastern Region and have not been brought to the main land and therefore, most of the people of India are deprived of such delicacies. If these plants are grown on a commercial scale and sent to Delhi and other metropolitan towns they will bring much benefit to the people of the North Eastern Region and people of other parts of India will have new delicacies on their table. The North Eastern region is also rich in ornamental plants which have not yet been explored on a commercial basis. These plants can change and bring a new look into the gardens. The orchids which are endemic to the area are very valuable. Arunachal Pradesh and Sikkim have exploited their potential and export orchid flowers to the tune of hundred crore rupees. This could be further developed and new hybrid plants produced by tissue culture augmenting the horticultural wealth of the country.

Many of the medicinal plants which are important for Ayurvedic preparation are also valuable foreign exchange earners. The details about these plants will not be discussed in this book. The details can be perused in another volume entitled "Medicinal Plants of North Eastern Region" by the author which is under publication.

Perilla frutescence which grows abundantly in the North Eastern Region, is a very useful plant. The oil obtained from the seed is used as cooking medium by the tribals, however this oil has a special quality which has not been exploited in this country. Japanese use this oil as seasoning for meat. The oil is also used for preparation of special polish which gives very good gloss. There are other products which can be prepared from this oil and have a great economic value. As already mentioned, these aspects have neither been explored nor exploited. The products of this oil can be a good foreign exchange earner. The total quantity of the oil available is about 15,000 tonnes. This being considerable amount can help in establishing paint and edible oil industry in the area. The oil of another plant *Butea minor* was sent to Harcourt Butler Technological Institute, Kanpur for chemical evaluation. They have reported that this oil can be used for the manufacture of paints and high quality soaps. It is rather a travesty that such valuable resources available in the region have not been exploited which could bring economic benefits by establishing paint and soap industries. It is, therefore, necessary that the potential of economic plant wealth available in this region should be fully understood and used for the benefits of the country which is reeling under dire poverty.

In addition to certain specific items mentioned above, these plants may also provide some other minor products about which we have practically no knowledge.

Almost 462 plants have been studied and they belong to the following 8 groups.

1. Dye yielding plants
2. Fibre yielding plants
3. Spices and condiments
4. Essential oil producing plants
5. Ethnotoxic plants
6. Gum, resin and tannin yielding plants
7. Timber plants
8. Food and Edible plants

The above group of plants are important for establishing agro-based industries as well as other industries. Also some of them may have medicinal value. The details are discussed under each head.

1. Dye Yielding Plants

There is now demand from all over the world that textiles be dyed with natural dyes as AZ0 dyes have been reported to be carcinogenic in the long run. India is one of the important textile exporters in the world and therefore it is necessary that sources of natural dyes be discovered to maintain India's supremacy in the world textile market. The export of textiles also earns valuable foreign exchange for the country. We can learn from the inhabitants from N.E. Region the technology of dying as they have been dyeing their cloth with natural dyes from times immemorial. This region thus can provide the resources of natural dyes to meet the national requirement and will also help increasing the employment potential. This will be an important source of income generation and thus improving the quality of life.

2. Fibre Yielding Plants

The same is true of this group of plants. The fibre plants available in this region have not been fully exploited. Some of the fibres can also be utilised in textile industry as admixture to improve the quality of cloth. The fibres of different types as discussed herein may be important import substitutes.

3. Spices and Condiments

Many of the spices and condiments can be easily grown in this area because of very suitable climate. This will help in the development of specialised agriculture and will bring far more benefits than growing agricultural crop. This will also meet the national requirement of the spices and may be available for export, earning valuable foreign exchange. Some of the spices are used locally and people on the mainland are deprived of these very useful spices. If these spices are brought to metro - towns then the people there will have new delicacies. This will help in the development of the region.

4. Essential Oil Producing Plants

The region gets its cooking oil requirement from the mainland. However, the tribals have been using a number of plants to meet their household requirement such as *Phosocarpus tetragonolobus*, *Perilla frutesence,* and *Butea minor.*

Many of the oils can be used for establishing soap and paint industries and many of the essential oils can form base for perfumes which will not only be exotic but will also develop perfume industry thus saving valuable foreign exchange spent on importing perfume bases.

5. Ethnotoxic Plants

Many of the plants can be source of insecticides of various types or may be found to be useful medicinal plants for dreaded disease like cancer and AIDS. However a detailed study is needed before we can draw definite conclusion. Some university in the N.E. Region can take up specific studies on these plants.

6. Gums, Resins and Tannin yielding Plants

These products are in great demand in the industry. Unfortunately these important resources have not been exploited to their full potential. The industrial houses as well as the government need to take interest in these valuable resources of the region.

7. Timber Plants

The N.E. Region is rich in various types of timbers both deciduous and evergreen. The N.E. Region meets more than 50% of the timber requirement of the country. However in recent years indiscriminate felling is disturbing the environment of the area causing irreparable harm to the region through floods, washing away useful soil crust and thus reducing the crop yields. This will result in catastrophe for the posterity. Some of the species may become endangered. There is, therefore, a need to use the timber plants with care and consideration. There is also need for establishing germ plasm nurseries to pervent extinction of useful species.

The germ plasm nurseries can also be utilized by research workers to breed fast growing timber plants to meet the country's need for timber.

Arunachal Pradesh forest department has taken a lead in establishing such a nursery. However, other N.E. States should also establish such nurseries.

8. Food and Edible Plants

Many of the food plants of the region are valuable and interesting. The variety is vast and needs to be conserved and studied in detail. The huge produce of these valuable exotic food plants cannot be consumed by local people and the mainland is deprived of these rare delicacies. If arrangements could be made to transport these plants to big Metro towns the people will be able to enjoy the delicacies and local people will benefit by increasing their earnings.

Many of these wild plants consumed by the people of N.E. Region are wild progenitors of crops, vegetables and food plants and may bestow disease and insect resistance as they survive under stress conditions.

The germ plasm of rice from this area has been used by International Rice Research Institute, Phillippines for producing useful varieties. The area has also the best citrus germ plasm of the world maintained at Tura (Meghalaya). It is unfortunate that we have not utilised this germ plasm for breeding useful varieties as well as new types of useful citrus plants.

This publication will lead to the following if the information is used properly :

A. Establish new industries.

B. Provide more employment to the inhabitants of the area and thus improve the quality of life.

C. It will increase developmental activities in the area removing regional imbalance.

D. It will result in saving valuable foreign exchange as many of the products obtained from these plants may be important import substitutes.

E. New items of food will be available for other parts of the country bringing integration and recognition of the importance of this remote and still unexplored region.

F. The studies on the germ plasm of various species will provide information about useful types.

G. It will improve overall environment of the area.

H. This publication will provide a catalogue of the useful plants and may prevent patenting of our useful plant wealth.

It will thus be observed that the present publication will not only provide the information about valuable plant wealth of the region but if properly utilised will bring all round development of the region, economic stability and better quality of life for the people.

2

DYE YIELDING PLANTS

1. Aporusa dioica Muell. - Arg. Euphorbiaceae

Common Name : Tamsir - arang

Distribution : Throughout the North Eastern region.

Small evergreen tree, Bark greyish brown, reticulately fissured. Leaves 6 - 15 x 2.75 cm, elliptic or elliptic lanceolate, acuminate at the apex, cuneate or rounded at the base, entire or obscurly crenate, glabrous sometimes hairy on the nerves beneath. Male spikes upto 5 cm long, often clustered, bracts broad, ciliate. Calyx pubescent. Ovary hairy, stigma short, bifid. Fruit nearly glabrous, ovoid oblong, beaked.

Uses : Leaf is utilized for preparation of black dye.

2. Baccaurea ramifolia (Roxb.) Muell.- Arg. Euphorbiaceae

(Syn.) *B. sapida* (Roxb.) Muell.- Arg.

Common Name : Dampiya - araung, Sohramdieng, Tampawk (AS).

Distribution : Throughout N.E. Region and also planted in homestead compounds.

Small or middle sized tree, young parts hairy. Bark dark grey, with vertical lenticels. Leaves 10-20 cm long, obovate, or elliptic obovate or oblanceolate, acuminate, glabrous, membranous. Flowers diocious, apetalous, shortly pedicelled in densely fascicled. Ovary tomentose. Male flowers subsessile, solitary or grouped on very small lateral branches of racemes in the axil of the bracts, calyx lobes 4. Stamens 4 - 8 free, anthers small pistillode 3 lobed. Female flowers yellow in simple racemes. Ovary tomentose, with two ovules in each cell, stigma peltate fimbriate. Fruit globose yellowish brown.

Uses : Leaves are utilised to produce a chocolate colour dye.

3. Bischofia javanica Bl. Euphorbiaceae

Common Name : Manipuri - Theichak, Tripura - Khungthi, Jakki, Phangput - araung.

Distribution : Manipur, Tripura and adjoining states. It is also found in shady ravines and river banks of sub - Himalayan forests of Punjab and moving eastward through Uttar Pradesh to West Bengal and Orissa.

Large deciduous tree with tall cylindrical trunk and a spreading crown. Bark dark grey. Leaves alternate, trifoliate, leaflets 8 - 15 x 3.5 cm ovate oblong or elliptic, terminal much longer, acuminate, crenate, glabrous. Flowers greenish, dioecious, apetalous in paniculate racemes. Male flower shortly pedicillate, Sepals 5, Stamens 5, filaments large, anther short, pistillode flat, discoid. Female flower longer than male, pedicelled. Sepals ovate, caducous. Ovary 3 - 4 celled with two ovules in each cell, Style linear, entire. Fruit globose, fleshy brown when ripe.

Uses : A red dye is obtained from the bark.

4. Bixa orellana Linn. Bixaceae

Common Name : Latakam, Powasi, Kungumapoovu, Ureirom.

Distribution : Commonly planted in house compounds throughout North Eastern region.

Small shrub or small evergreen tree about 4 m in height. Leaves alternate, simple, petiolate 8 - 20 x 5 -13 cm broadly ovate to cordate scaly beneath, very densely red dotted. Flowers white or pinkish. Sepals 4 - 5 free. Stamens numerous. Ovary superior, unilocular with two opposite parietal placenta.

This species occurs in two forms one with white flowers and green capsules and the other with pink flowers and red capsules. Each capsule contains about 50 small seeds, pulpy outer covering contains pigments.

Uses : A dye is obtained from seeds. For extracting the dye the seeds are bruised and the pulp macerated with hot water in wooden vessels and soaked for several days till the colouring matter forms a fine suspension. The seeds are then removed and the liquid which contains the pigment is allowed to ferment for about a week. The dry "Amatto" settled at the bottom is separated and dried into cakes.

5. Butea minor Buch. - Ham. Fabaceae

(Syn.) *B. bideiformis* (Voigt) Grier.

Common Name : Fu.

Distribution : Common in valley in open grassland, scruby jungles, by the sides of drains and in partially submerged area. It grows in association with *Vachellia farensiana*.

Erect shrub with long spreading branches covered with grey silky tomentum. Leaves trifoliate, petiole channelled, tomentose, leaflets oblique, large, grey silky beneath. Flowers in red racemes.

Uses : Flowers yield a red dye.

6. Carthamus tinctorius Linn. Asteraceae

Common Name : Colop - Machu, Gulandaud.

Distribution : Throughout N.E. Region.

Thistle like branching herb. Leaves sessile 1.3 cm- 1.5 cm, lanceolate, usually spinously serrate. Head large, terminal. Flower orange red. Involucre bract many, serrate, outer foliaceous, green, usually spinous, inner lanceolate, ovate, oblong acute. Fruit cypsela, obovoid, 4 - angled, pappus absent.

Uses : The dye is obtained from the flowers.

7. Curcuma caesia Roxb. Zingiberaceae

Common Name : Kal - Halda.

Distribution : Cultivated in gardens throughout N.E. Region.

A perennial herb. Rhizomes pale grey inside. Leaves large oblong glabrous 30 - 60 cm long having deep ferruginous purple strips in middle which continue to lower surface. Flowers pale yellow, reddish at the outer border, shorter than bracts.

Uses : Rhizomes produce black dye.

8. Flemingia stricta Roxb. Fabaceae

(Syn.) : *F. macrophylla* (Willd.) Merr.

Common Name : Charchra, Bhanjabasa, Charariaw.

Distribution : Throughout N.E. Region at lower elevation in shady and more or less damp localities and ravines in hills and valleys.

Shrub. Leaves trifoliate, leaflets 15 - 30 cm long, oblong narrowed at both ends, glabrous above, pubescent below. Racemes 5 - 10 cm long. Calyx teeth linear, lowest longest. Calyx, toothed. Corolla as long as the calyx, Pod glabrescent. Seeds black, deltoid compressed.

Uses : Pod hairs yield a purple dye.

9. Garuga pinnata Roxb. Burseraceae

Common Name : Kakad, Tejhi, Timur - araung, Tejhi.

Distribution : Throughout N.E. Region.

Deciduous middle sized tree. Leaves alternate, exstipulate, imparipinnate, leaflets 11 - 19 cm. Opposite subsessile oblong or ovate, lanceolate, caudate, acuminate, crenate, subcoriaceous, pubescent beneath. Flowers greenish yellow in cymes, arranged in large terminal panicle. Calyx campanulate. Petals 5, yellow about thrice as long as calyx. Stamens 10 equal in two series. Ovary sessile, ovoid, 4 - 5 celled, style hairy. Stigma capitate, 4 - 5 lobed.

Uses : The fruit is edible. The leaf galls are used for tanning. Bark gives black dye.

10. Indigofera dosua Buch.-Ham. ex D. Don. Fabaceae

Common Name : Dien - sha - ratoh, Bujir, Bilhi, Buthi, (Dule).

Distribution : Throughout N.E. Region up to 2000 m fairly common in grassland.

A densely hairy or tomentose shrub. Leaves subsessile, leaflet opposite, many, linear, oblong to elliptic oblong, mucronate, densely pubescent. Racemes densely many flowered. Bracts lanceolate, cuspidate. Corolla pink, purple.

Uses : Leaves give a deep blue dye.

11. Indigofera tinctoria Linn. Fabaceae

Common Name : Gali, Jarmooli, Kiron, Nil, Tarhalim.

Distribution : Throughout N.E. Region.

A shrub up to 2 m high. Leaves compound 2.5 - 5 cm long, leaflets opposite, membranous, turning blackish when dry, leaflets 9 - 13, obovate - oblong. Racemes lax, nearly sessile, 5 - 10 cm. Corolla reddish - yellow. Pod nearly straight 2 - 2.5 cm long.

Uses : Leaves give a blue dye.

12. Kydia calycina Roxb. Malvaceae

Common Name : Bothi, Moti hirvanti, Varnag, Vellasaclachi.

Distribution : Throughout N.E. India.

Tree 5 - 10 m tall. Leaves ovate to suborbicular, cordate to truncate at base, palmately 3 - 5 lobed at apex, 6 - 16 cm diam, palmately 5 - 7 nerved, pubescent beneath, petioles 3 - 12 cm. Flower white, 12 - 8 cm across, in terminal panicles. Bracteoles 4 - 6 oblong, hispid. Capsules depressed, globose.

Uses : The lac of this plant is used as dye for colouring clothes etc.

13. Lithocarpus fenestratus (Roxb.) Rehder Fagaceae

(Syn.) *Quercus fenestrata* Roxb.

Common Name : Kashiendung (Lepcha); Kuhi, kadieng jing (Assam & Manipur).

Distribution : Entire N.E. Region in the plains and forms scrub forest at the edges of the hills in and around the valley.

Middle sized large evergreen tree. Leaves 10-17.5 x 2.5-6 cm oblong, lanceolate, elliptic, lanceolate, acuminate, entire coriaceous. Inflorescence tomentose, crowded at the end of branches in panicles bearing both male and female flowers. Male flowers in glomerules of 2-3, bracteolate. Perianth segments, 5-6, tomentose outside. Female flowers in glomerules of 3, bracteolate.

Uses : Nuts provide orange dye. The bark and acorns are reported to be used for dyeing in Manipur. The bark contains 10 - 16% tannin and can be employed as a tan stuff. It gives a leather of light colour.

14. Macrosolen cochinchinensis (Lour.) Tiegh Loranthaceae

Common Name : Chakjo - ki - akuldo.

Distribution : Throughout N.E. Region.

Shrub, semiparasitic. Leaves opposite, rarely alternate, ovate, elliptic, lanceolate, sub acuminate at the apex, acute or cuneate at the base, coriaceous, glabrous. Flowers about 1.5 cm long in axillary racemes. Bracts ovate. Calyx truncate, Corolla greenish orange, tube campanulate. Fruit globose or ellipsoid, pale yellow.

Uses : Wood gives a yellow dye. The plant contains quercitrin and a wax.

15. Mahonia manipurensis Takeda Berberidaceae

Common Name : Unapu

Distribution : Entire N.E. Region up to 2100 m.

Shrub with compound leaves, Leaflets 5 - 9 pairs ovate. Racemes in fascicles of 5 - 8 flowers.

Uses : Yellow colour is extracted from the bark. It also gives alkaloids - oxyacanthine, berberin, neprotine and Jatrorrtine. *M. nepalensis* also, gives a yellow dye which is extracted from roots.

16. Mallotus philippinensis (Lamk.) Muell. - Arg. Euphorbiaceae

Common Name : Jhapad, Komada, Kunmal, Rohini, Sindur.

Distribution : Entire N.E. Region.

Small evergreen tree. Leaves alternate 7.5 - 15 cm long, ovate - oblong or lanceolate, acuminate, entire or dentate, glabrous above, pubescent below. Flowers small dioecious. Male flowers sessile or so, clustered in erect terminal spikes of panicles. Sepals 4, stamens numerous. Female flowers sessile in short spike. Capsules 0.8 - 1.2 cm across 3 lobed and 3 valved covered with bright red powder composed of fine grains of a resinous substance mixed with minute stellate hairs.

Uses : A red dye is obtained from the fruits which is used for colouring textiles particularly silk and wool. The colouring matter is present in the red glandular hairs.

17. Mammea suriga (Buch.-Ham ex. Roxb.) Kosterm. Clusiaceae

Common Name : Suklong, Lushai, Urdi.

Distribution : Throughout the N.E. Region.

Small or large tree with dense deep green foliage, young shoots obscurely angled. Bark greyish brown, inside reddish, turning brown, exuding milky sap. Leaves 17.5 cm/ 5 cm oblong lanceolate, usually acute or bluntly pointed, coriaceous, quite glabrous, midrib prominent veins elegantly reticulate, base obtuse, petiole stout. Flowers 5 cm across in dense short, fascicles in axils of fallen leaves. Buds globose orange red. Calyx reflexed during flowering. Petals white streaked with red. Stamens sterile in female flowers, style short. Fruit obliquely ovoid.

Uses : The flower buds are used for dyeing silk.

18. Morinda angustifolia Roxb. Rubiaceae

Common Name : Tarlaung - araung Chenong.

Distribution : Frequently found at the outskirts of forests throughout N.E. Region.

Small tree. Leaves simple, opposite, glabrous, large, elliptic or obovate - lanceolate, caudate, acuminate, tapering into petiole. Flowers fragrant in axillary umbels, in globose heads, ultimately elongating in fruit.

Uses : The root is used for dyeing cotton and also wool and silk. *M. citrifolia* and *M. tomentosa* are considered chief source of all dyes. *M. bracteata* and *M. umbellata* are also exploited for the same purpose.

19. Mussaenda roxburghii HK. f. Rubiaceae

Common Name : Sapaklari - naro, Hanulei.

Distribution : Entire N.E. Region.

Shrub. Leaves simple, opposite, short petioled, oblong - lanceolate, glabrous. Flowers orange, occasionally yellow in dense corymbose cymes. Calyx deciduous. Calyx segment becomes more or less enlarged and white. Corolla tube pale green.

Uses : An infusion of the leaves is used for colouring baskets.

20. Nyctanthes arbor-tristis Linn. Oleaceae

Common Name : Sephalika, Singgvalei, Harsingar, Khatkhosa, Shihor.

Distribution : Throughout India.

Small deciduous tree or bushy shrub. Branchlets rough pubescent 4 angular, bark rough, brown, greyish or greenish. Leaves 13.75-7.5 cm, ovate, shortly acuminate, distantly toothed or entire, coriaceous, scabrid, lateral nerves 4-6 on either half. Flowers sweet scented, sessile, 3-7 together on hairy quadrangular peduncles of various length which are arranged in cymes or cymose panicles, Bracts elliptical, pubescent. Fruit capsule. Seed exalbuminous.

Uses : Flowers are used for dyeing.

21. Onosma emodi Wall. Boraginaceae

Common Name : Shankhuli.

Distribution : Throughout N.E. India.

Hispid herb. Leaves sessile, lanceolate or hispid. Flowers in capitate cymes, pedicels 0.75 cm long. Sepal 0.75 cm long, ovate - acuminate, hispid, spreading in fruit. Corolla purplish ventricose narrowed at the mouth, hairy, 1.5 cm long. Anthers inserted, style exserted. Nutlets 4, tuberculate.

Uses : Roots yield a dye, which is used for colouring wool and silk.

22. Oroxylum indicum L. Vent. Bignoniaceae

Common Name : Napak - banaraung, Tappakaya, Samba, Tetoli, Venga, Toguna, Bhatghila, Dingari.

Distribution : Common throughout N.E. Region.

Small tree with few branches and open crown. Leaves opposite, ternate or bipinnate 1 - 2 m long. Leaflets broad - ovate entire acuminate, glabrous. Flowers large fleshy purplish in long terminal lax racemes. Calyx fleshy or coriaceous. Corolla campanulate. Stamens 5. Ovary compressed. Fruits long, sword shaped. Ovules numerous.

Uses : Bark and fruit are used in tanning and dyeing.

23. Pithecellobium angulatum Benth. Mimosaceae

(Syn.) : *P. heterophyllum* (Roxb.) Haridasan & Rao

Common Name : Theng - Semai - araung. Inghane, Jangal - jalebi, Kodukapully.

Distribution : Common in the entire N.E. Region.

Small or middle sized tree. Leaves bipinnate, rachis angled with a large gland at the base, pinnae 2 - 4 pairs. Leaflets 5 - 10 pairs in upper pinnae, 2 - 3 pairs in lower pinnae, oblong lanceolate to elliptic or rhomboid, ovate. Flowers yellowish white in peduncled umbellate corymbs. Calyx campanulate. Corolla pubérulous outside, ovary puberulous, short stalked.

Uses : The fruit and leaves are used for dyeing along with *Terminalia chebula* which gives a black dye. The plant is also grown in gardens as an ornamental plant.

24. Rubia cordifolia Linn. Rubiaceae

Common Name : "Tsu" or "Tot", Bad - rahoi, Iratiga, Manjishtha, Saphrang.

Distribution : Throughout North East India.

Climber. Leaves opposite, often with 2 foliaceous stipules giving the appearance of whorled nature, ovate - lanceolate, rounded or subcordate at base, long - acuminate, 3 - 10 x 1.5 cm, nerves 3 - 5 palmate, petioles 4 - 8 cm, panicles much branched, 3 - 15 cm, pedicels 1 - 2 mm. Corolla green or orange red, lobes spreading. Fruits globose, purplish - black.

Uses : Root and stem of the plant are used for preparation of dye.

25. Smilax ocreata A. DC. Smilacaceae

Common Name : Plalangtung rikang.

Distribution : Entire N.E. Region.

It is a climber with prickly stem and smooth branches.

Uses : Fruit gives a yellow dye.

26. Strobilanthes flaccidifolius Nees Acanthaceae

Common Name : Rum, Rampat, Raspat.

Distribution : Entire N.E. Region.

Glabrous shrub. Leaves elliptic acute at both ends. Flowers in densely lax spikes. Bracts petioled, ovate. deciduous. Calyx linear, spathulate. Corolla long glabrous.

Uses : The plant is cultivated for blue dye and three crops are obtained in a year.

27. Symplocos paniculata (Thunb.) Miq. Symplocaceae

Common Name : Dieng - long, Lodh.

Distribution : Entire N.E. Region between 600 - 2400 m.

Deciduous shrub or tree upto 12 m high. Leaves obovate, softly pilose beneath. Flowers white, fragrant in terminal panicles. Fruits blue, drupaceous.

Uses : The bark and leaves yield a yellow dye.

28. Woodfordia fruticosa Kurz. Lythraceae

Syn. : *Woodfordia floribunda* Salisb.

Common Name : Dheira, Dhai, Dhavai, Icha-Bahor, Sinjanape.

Distribution : Throughout the N.E. India.

Large spreading shrub with reddish brown bark peeling off in thin strips, blaze pink, young parts hairy. Leaves and branchlets covered with black dots. Leaves sessile or sub sessile, opposite, sometimes in whorls of three, distichous, lanceolate or ovate - lanceolate, acuminate, puberulous above. Flowers scarlet, shortly pedicellate, zygomorphic, in short panicles of closely clustered cymes. Calyx scarlet, long, persistent tubular, somewhat curved, oblique at the mouth. Petals 6, white, acute. Stamens 12, inserted at the base of the calyx tube. Ovary superior 2 - celled. Ovules numerous. Seed numerous, minute.

Uses : The flowers give reddish or yellowish dye which is used for dyeing silk. The leaves and flowers also contain a large percentage of tannin.

3

FIBRE YIELDING PLANTS

29. Abelmoschus moschatus (L.) Medic. Malvaceae

Common Name : Tachusenghme, Dhokrakanda, Gorukhia-korai.

Distribution : Throughout N.E. India.

Annual, erect, hispid herb or under shrub up to 1.5 m tall. Leaves variable orbicular elliptic 6 - 15 cm long, base cordate, 3 - 7 lobed. Flower axillary solitary. Epicalyx segment linear,calyx up to 3 cm long, 5 toothed, stellate tomentose outside. Petals 6 - 8 cm long obovate yellow with purple base. Capsule 3 - 6 cm long, ovoid oblong, seeds 3 - 5 mm long sub - reniform, ribbed.

Uses : The raw cotton obtained from the fruit is used for weaving.

30. Abroma augusta Linn. Sterculiaceae

Common Name : Miri - Ulat Kambal, Bon - Kopahi.

Distribution : In scrub forest of Assam ascending up to 1000m.

A large shrub or small tree with downy branches. Leaves 1.0 - 2.5 cm long, narrow to the 3 - 7 nerved base, repand denticulate, glabrescent above, tomentose below, stipules linear, deciduous, as long as the petiole. Calyx lobed, lanceolate, free nearly up to the base, petals slightly exceeding the sepals, imbricate in buds, deciduous. Capsules 2.5 - 5.0 cm in diameter, persistent calyx, glabrous or nearly so when ripe.

Uses : The bark of the twigs yields a strong silky fibre which may be used for a variety of purposes and for which the plant is well worth cultivating.

The fibre is readily separated by retting the bark in water for 4 - 8 days. It is pale yellow in colour, soft and glossy. It makes a pliable and an attractive rope which compares favourably with machine made abaca rope. It resembles jute in chemical behaviour and composition.

31. Abutilon indicum (L.) Sweet Malvaceae

Common Name : Potari, Jhapa, Kanghi, Kanski.

Distribution : Throughout N.E. Region.

A perennial herb, sometimes up to 2.5 m in height. Stem covered with soft velvety tomentum. Leaves 5 - 6, nearly as long as broad, ovate orbicular, cordate, acute or acuminate, entire or distantly irregularly toothed, pubescent above, grey felted and also with scattered hairs underneath. Flowers golden yellow.

Uses : The stem is retted and yields good fibre which is suitable for ropes.

32. Anaphalis busua (Buch. - Ham. ex D. Don) DC. Asteraceae

Common Name : "Tiew Riem"

Distribution : Throughout N.E. India.

Shrub erect, usually much branched, leaf base hairy. Basal leaves inversely lance - shaped. Stem leaves linear 1.5 cm sharply pointed, margin often bent back, basal lobes usually long, sharp pointed and decurrent. Flowers many in corymb.

Uses : The stem is used for filling pillows.

33. Artocarpus chaplasha Roxb. Moraceae

(Syn.) *Artocarpus chama* Buch. - Ham. ex Wallich.

Common Name : Cham, Herukok Tong, Tatpong, Pram, Arma, Dahua, Ingtal, Chaplash.

Distribution : Meghalaya, Manipur, Arunachal Pradesh.

A tall deciduous tree. Leaves alternate scabrid elliptic ovate stipules amplexicaul. Flowers monoecious, densely crowded on globose axillary receptacle. The greyish brown bark exudes a milky juice.

Uses : Stem and bark yield valuable fibre which is used for making blankets.

34. Artocarpus lacucha Buch. Ham. Moraceae

(Syn.) *Artocarpus lakoocha*

Common Name : Armu, Burhal, Delephal, Ingtal.

Distribution : Assam.

A deciduous tree with a clear straight stem. The tree is also cultivated for its edible fruits.

Uses : Stem bark yields fibre which is used for making crude cordage.

35. Arundinaria callosa Nakai Poaceae

Common Name : Apa = Tabto, Ad = Bud = Biyi.

Distribution : Throughout N.E. Region.

A shrubby thorny bamboo, culms 4 - 7 m greyish green, smooth striate when young, nodes raised in a ring which is subtended by the base of the fallen sheath fringed with soft brown hairs which is studded with thick short conical spines, the spines breaking through the sheath when the latter is detached. Leaves oblong - lanceolate, thin, unequally attenuate at the base, short petiole, terminating in a scabrous acuminate point, finally grandular. Inflorescence a branched panicle ending in leafy branchlets.

Uses : Bark is used for making ropes. The split culms of most of these bamboos are locally employed for cottages and for making mats and baskets.

36. Bauhinia vahlii (Wt. & Arn.) Benth. Ceasalpiniaceae

Common Name : Bhorla, Sungungrik, Chehur, Malghain.

Distribution : Throughout the N.E. India.

A liana often ending in a pair of tendrils, young parts, petioles and leaves, specially along the nerves, densely rusty or rarely greyish tomentose. Leaves very variable in size, usually 11.25 - 47 cm across, sometimes smaller, as broad as long or broader, lobes obtuse, rounded. Flowers white or cream coloured in terminal woolly corymbs or corymbose racemes. Fruit - pod.

Uses : The bark yields strong and variable fibre used for making cordage and ropes.

37. Boehmeria macrophylla D. Don Urticaceae

Common Name : Kamli, Pua.

Distribution : Throughout N.E. Region.

A small tree or a large shrub with branchlet quadrangular and young parts strigose. Leaves petiolate opposite,ovate to lanceolate, acuminate, coarsely toothed, scaberulous, 3 nerved. Flowers monoecious in dense clusters arranged in drooping racemes. Male flowers : perianth 1 lobed, stamens 4, pistillode rudimentary. Female flower : perianth 2 - 4 toothed, persistent, styles laterally stigmatose. Achene closely invested by perianth.

Uses : Stem is used for preparing fishing nets and ropes.

38. Boehmeria malabarica Wedd. Urticaceae

Common Name : Tak briet - Kung, Sam - khul - khia, Borthar thari.

Distribution : Throughout N.E. Region.

An erect shrub or a small tree, branches spreading. Bark greyish, rough thin. Leaves sub - equal or alternately large and small, ovate, ovate - lanceolate or elliptic - lanceolate, crenate or serrulate, acuminate, membranous, glabrous pubescent beneath, strongly three nerved. Flowers greenish white in axillary sessile clusters. Male flowers : perianth four partite stamens four, pistillode small, sub - globose. Female flowers : perianth four toothed, pubescent. Fruits achene, minute.

Uses : Fibre is used for making ropes.

39. Boehmeria nivea Gaud. Urticaceae

Common Name : Riha

Distribution : Throughout N.E. Region.

A shrub, branches tomentose. Leaves 6.75 - 3.75 cm broadly ovate or sub - orbicular, acuminate, dentate, sub - coriaceous, scabrid above, white beneath, densely matted with adpressed hairs. Flowers exserted, hairy. Fruit achene, oblong, compressed hairy.

Uses : Bark yields fibre which is used for weaving clothes,bags and ropes. The fibre is regarded as the longest and most silky of all vegetable fibres. It has great strength and durability and is highly resistant to the action of water. Gas mantle of good quality are also made of it.

40. Boehmeria platyphylla D. Don Urticaceae

Common Name : Kamli, Taksur, Samrali.

Distribution : Throughout N.E. Region.

A variable large spreading shrub, branches succulent more or less strigose. Bark blackish, contains a pith like substance inside. Leaves opposite,sometimes alternate, broadly ovate, ovate elliptical or sub - orbicular, acuminate, coarsely toothed rather membranous, more or less scaberulous, three nerved, base rounded or cordate, often oblique. Flowers whitish, monoecious. Male spike usually branching from lower axils. Female spike from upper axils fruiting, perianth ellipsoid, sometime angled.

Uses : Bark yields a strong fibre.

41. Chimonobambusa intermedia Nakai Poaceae

Common Name : Tite - Nigalo, Perinyok.

Distribution : Throughout N.E. Region.

A slender caespitose bamboo, culms smooth greenish green 3 - 4 m tall, nodes swollen with a prominent ring and a whitish line below them, internodes 12.5 - 25 m long, branches many from the nodes. Leaves bright green, very variable, attenuate at the base into a short swollen petiole, terminating in a setaceous point. Inflorescence on a leafless stem consisting of racemose panicles, verticillate from the nodes, subtended by membranous bract.

Uses : Stem is used for making baskets and preparation of ropes.

42. Cissampelos pareira Linn. Menispermaceae

Common Name : Pctuepati, Tumshupripk, Akanbindi, Barawent.

Distribution : Throughout N.E. Region.

A climber which grows annually from perennial root stocks, young part usually tomentose or pubescent. Leaves orbicular, peltate, obtuse or sometimes retuse, mucronate at the apex, membranous or subcoriaceous, pubescent, grey tomentose beneath or both surface glabrate specially the upper, base cordate or truncate, 5-7 nerved, petiole as long as, or longer than the blade, tomentose or glabrate. Male flowers in cyme. Female flowers racemed and crowded in the axil of leaf like bract.

Uses : It yields a strong fibre for making ropes.

43. Corchorus capsularis Linn. Tiliaceae

Common Name : Thetamurapath, Tita, Jute.

Distribution : Throughout N.E. Region.

An annual shrub. Leaves oblong, acuminate, coarsely toothed, base prolonged into tail like appendages. Flowers small yellow, bracteate, Sepals 4-5. Petals 4-5. Stamens generally numerous on a short torus. Ovary 2-5 celled, ovules, many in each cell, style short, Capsule oblong not beaked, wrinkled, muricate 2-5 valved without transverse septa, seeds numerous.

Uses : The bark gives jute fibre.

44. Corchorus olitorius Linn. Tiliaceae

Common Name : Bampat, Naharacha.

Distribution : Throughout N.E. Region.

An annual shrub. Leaves oblong, acuminate, coarsely toothed, base prolonged into tail like appendages. Flowers small yellow, bracteate. Sepals 4-5. Petals 4-5. Stamens generally numerous on a short torus. Capsule elongate, glabrous, ribbed, beaked, cells transversely septate.

Uses : It is considered to be an indigenous plant which gives good fibre for ropes.

45. Daphne papyracea Wallich ex. Steud Thymelaeaceae

Common Name : Kagate, Argayle.

Distribution : All Assam and other N.E. Region including Manipur up to a height of 2100-2400 m.

A shrub attaining a height of 2-3 m. Leaves lanceolate to narrow lanceolate, acuminate or acute, entire, coriaceous, glabrous, dark green above pale beneath. Flowers pale yellow to purple or white, scented crowded in head like clusters or very short peduncled, 6 filaments very short, ovary on a minute sub - tracte stipe which is narrowed towards the base. Style very minute, Stigma capitate. Fruits orange, black.

Uses : The bark which can be easily peeled off from the shrubs is employed for making paper in the region.

46. Debregeasia longifolia (Burm. f.) Wedd. Urticaceae

Common Name : Tusare, Tutomtanam, Jirepol, Kapsi, Tushiaru.

Distribution : Assam.

A small ornamental tree or shrub, spreading type 7 to 8m in height. 0.75 to 1 m in girth. Young shoots silky pubescent. Leaves oblong, lanceolate or elliptic, acuminate, serrulate, sub - coriaceous, scabrid above, pale white or ashy tomentose beneath. Flowers in lateral dichotomous panicles. Male flowers, bracteate sub-orbicular. Sepals exceeding the bracteoles, pistilloid glabrous. Female flowers bracteate narrower. Fruits an aggregate of achenes in clusters. The fruit is edible.

Uses : The plant is a source of fibre and is of superior quality and is valued for ropes and cordages.

47. Dendrocnide sinuata (Blume) Chew. Urticaceae

Common Name : Moriange, Ongyalap, Gilmatjokma, Utigan.

Distribution : Throughout N.E. Region.

Shrub with stinging hairs. Leaves ovate or elliptic, more or less entire, sometimes crenulate towards the apex, acute or acuminate, clothed with stinging hairs. Flowers in cluster. Fruits inflated white.

Uses : It yields a fibre, used for making ropes.

48. Dillenia pentagyna Roxb. Dilleniaceae

Common Name : Aggai, Kallai, Akashi, Karkat, Karmatta.

Distribution : It is common in sal forests of Assam and Bengal and is scattered throughout ravines and hill slopes in west Manipur.

A large deciduous tree with straight cylindrical pole 9 - 21 m in height, 2.5 m in girth with ascending branches bearing a rounded crown of large leaves. Leaves oblanceolate, very narrowed at the base, often stem clasping, serrate, coriaceous and hard, glabrous above and glabrescent beneath. Flowers fragrant, yellow 3 cm across. Berries in umbels. Fruit orange yellow succulent and edible.

Uses : It yields a fibre for cordage. The leaves are used for wrapping and thatching where grass in scarce.

49. Edgeworthia gardneri (Wallich) Meissn. Thymelaeaceae

Common Name : Argayle, Kaigate, Dhenak, Phakrawon.

Distribution : Sikkim, Nagaland, Manipur.

A large shrub, branches long, scrambling. Leaves oblong elliptic or elliptic lanceolate, acute acuminate, glabrous above somewhat pilose beneath, base acute. Peduncle simple ebracteate. Flowers yellow in dense hemispherical heads. Perianth sub - villose, tomentose, lobes ovate, obtuse. Stamens 8, biseriate. Receptacle tubercular silky, ovary villose. Fruit included in the base of the perianth, pericarp coriaceous, testa crustaceous.

Uses : Stem bark yields valuable fibre for making ropes.

50. Edgeworthia tomentosa (Thunb.) Nakai Thymelaeaceae

Common Name :Thouthua, Baramp thurisopa, Balom turi.

Distribution : Throughout N.E. Region up to a height of 2100m.

A large shrub with long scrambling branches. Leaves oblong, elliptic or elliptic lanceolate. Flowers yellow in dense, hemispherical heads.

Uses : The fibrous inner bark is a valuable material for paper. The pulp is obtained by cooking bark material in lime soda which gives white paper. This species is also cultivated in Japan and used in the manufacture of high grade paper for currency notes.

51. Ficus rumphii Bl. Moraceae

Common Name : Pair

Distribution : Himalayan tract ascending to 1500 - 2000 m particularly in Garo hills and Manipur.

A medium sized large deciduous tree with 3 - 6 oblong ovate or broadly ovate acuminate leaves sub - coriaceous, quite glabrous, upper surface dotted, and shining. Male flowers few near the mouth of the receptacle, perianth segments spathulate. stamen 1. Female flowers, perianth segments 3 lanceolate, ovary ovoid, smooth, style elongate, stigma clavate, receptacle sessile in axillary pairs, globose black when ripe.

Uses : The bark is greyish smooth inside and is used for cordage.

52. Girardinia palmata (Forsk.) Gaud. Urticaceae

Common Name : Simuee, Kazoobee.

Distribution : Throughout N.E. India.

A tall stout coarse soft stemmed under shrub, armed with numerous slender stinging hairs. Leaves dentate or serrate, membranous, pubescent, 3 nerved, base cordate. Male flowers 4 numerous in short axillary spikes, Velaet Petals, filament persistent, pistillode globose. Female flowers in compact capitate panicles, perianth tubular, ventricose, 2-3 toothed ultimately splitting, spathaceous. Fruit achene.

Uses : It yields a fibre used for making ropes, twine and coarse cloth. The fibre is present in the inner bark of the stalk.

53. Grewia disperma Drummond Tiliaceae

Common Name : Diengdohthli, Bolmengu.

Distribution : All over N.E. Region ascending to 1000 m particularly Meghalaya.

A small tree but sometimes 15 m high, young shoots glabrous or nearly so. Bark thin dark grey brown outside, uniformly coloured inside. Leaves oblong, lanceolate, acuminate, acute. Flowers axillary cymes. Fruit drupe.

Uses : Bark yields a strong fibre suitable for ropes and cordages.

54. Grewia hirsuta Vahl Tiliaceae

Common Name : Huktapata, Sohsinthing, Chatarni, Dhamni.

Distribution : Throughout in the plains of N.E. Region in grassland and forest.

Small spreading shrub with slender stellately pubescent branches. Leaves ovate or lanceolate or oblong, acute or acuminate, irregularly serrate, heavy and scabrous, densely tomentose beneath, 3 nerved at the base, polygamous, peduncle 1 - 3, axillary, 1 - 4 flowered. Drupes 4 lobed, brown, hirsute and shining.

Uses : The stem bark is used for making fibre and mats.

55. Grewia sapida Roxb. Tiliaceae

Common Name : Thaurogate, Phuhura, Barbe, Burso.

Distribution : Throughout N.E. India.

A prostrate undershrub with perennial woody rootstock throwing out annually a number of switching shoots which after producing leaves, flowers, fruits, are burnt or browsed down to be similarly renewed next year. Leaves ovate, obovate - oblong or orbicular, usually serrate, apex rounded or broadly acute. Flowers yellow, 2-5 in each peduncle. Sepals reddish brown inside. Fruit drupe edible.

Uses : The bark yields a strong fibre suitable for cordages.

56. Grewia sclerophylla Roxb. ex G. Don Tiliaceae

Common Name : Pharsia, Taglar, Gaphri.

Distribution : Throughout N.E. India.

Shrub or undershrub, young parts covered with densely brown stellate tomentum. Leaves 8-11 cm long, ovate to sub - orbicular, irregularly serrate, scabrous on upper surface, densely stellate, tomentose beneath. Base 3-5 nerved. Flowers white. Drupe, globose rough, hairy.

Uses : Stem yields a good fibre used for making ropes.

57. Hibiscus macrophyllus Roxb. ex Hornem. Malvaceae

Common Name : Pohundal, Maomarli, Tyllendkhar.

Distribution : Assam, Meghalaya, Nagaland.

A small deciduous tree, young shoots, petioles, stipules and inflorescence shaggy with long tufted yellowish brown bristles up to 1 m long and stellate tomentum. Bark greyish white some what rough on old stems, blase at first yellowish brown with streakes of brown tissue, turning darker brown after exposure. Leaves orbicular, deeply cordate, abruptly short, acuminate with entire or minutely crenulate on luxuriant shoots, thinly coriaceous intricately stellate, tomentose on both surface especially the lower, palmately 7 - 9 nerved turning yellow brown before falling. Flower in axillary and terminal cymes, supported by a pair of large spathaceous caducous bracts. Fruit capsule. Seed bearded with shiny hairs.

Uses : The inner layer of stem bark yields fibre suitable for cordage.

58. Hibiscus sabdariffa L. Malvaceae

Common Name : Patwa, Lalmista, Chukiar, Mesta, Ambla - Kundrum.

Distribution : Throughout N.E. India.

An annual erect shrub, with red or green stem, particularly unbranched or with branches near the base of the stem, glabrous or slightly hairy. Leaves variable, entire or lobed, glandular beneath. Flowers large, yellow with dark crimson eye. Epicalyx united at the base and adnate to calyx. Calyx connate below, free above, dark purple fleshy. Capsule ovoid, pointed, villose. Seed numerous, large reniformed, black brown, covered with minute stellate hairs.

Uses : The inner bark yields a fibre used locally for coarse ropes. The fibre is strong when green but becomes brittle on drying. Ropes made from the bark fibre are used for tying rafts.

59. Kydia calycina Roxb. Malvaceae

Common Name : Kubuide, Dansasival, Bothi, Moti hirvanti, Varang.

Distribution : It is common in sub - Himalayan tract mixed with deciduous forests in dry areas in N.E. Region.

A moderately sized fast growing deciduous tree up to 10 - 15 m in height, approximately 1 m in girth. Bark grey exfoliating in irregular flakes or thin long streakes. Leaves alternate, sub - orbicular often broader than long, palmately 5 - 7 nerved. Closely felted beneath. Flowers white or purplish in much branched panicle. Capsule small, globose. Seeds reniform, dark brown.

Uses : Bark yields a strong fibre used for tying and making basket. The wood yields a mechanical pulp of light shade but poor strength due to the short length of ultimate fibre. The pulp can be used as an admixture with 30% bamboo pulp for news print.

60. Maoutia puya Wedd. Urticaceae

Common Name : Pua, Keeongabee, Ja - khi, Bad - langmyrsau.

Distribution : Throughout N.E. India.

A shrub, bark purplish grey with vertical lines of brown lenticels, young parts hairy. Flowers minute, monoecious or dioecious in small cymose globose head dichotomously panicled sometime endrogynous. Male flowers : sepals 5 velvet, stamens 5, inflexed in bud, pistillode woolly. Female flowers : perianth ovary straight, stigma pedicillate, ovule erect. Fruits achene very small with fleshy pericarp.

Uses : It yields white strong fibre used for making fish nets, twine and clothes. Bark yields a strong fibre called puya or Nepal hemp resembling ramie in general characteristic. The plants are cut when the seed is just being formed to obtain the fibre. The bark is peeled off, and dried in sun for a few days and boiled with wood ash for 5 - 6 hours. The fibre is separated by beating with wooden mallets and then washed with water and dabbed with micaceous clay, dried and finally cleaned from clay and bark by rinsing with water.

61. Marsdenia tenacissima Wight & Arn. Asclepiadaceae

Common Name : Sunamurai.

Distribution : Throughout N.E. India.

A large twining shrub, extremities under surface of leaves and inflorescence, softly tomentose, broadly ovate or orbicular, acuminate, sub - coriaceous, velvety. Flowers greenish yellow, in much branched corymbose cymes, peduncles intrapetiolar. Fruits follicles.

Uses : The fibre is strong and is used for fishing lines etc. The stem is cut into section and the bark stripped to obtain the fibre. It is then dried, steeped in water for a brief period and then scraped and cleaned of adherent material. The fibre thus obtained is elastic and resistant to alkali, it is considered second only to rhea (*Boehmeria nivea*).

62. Mastersia assamica Benth. Fabaceae

Common Name : Rem.

Distribution : Arunachal Pradesh.

A stout woody climber, branches sarmentose, lenticelled, twining anti-clockwise. Leaves pinnately 3 foliate, common petiole stipule broader at the base than long. Leaflets ovate or obovate, tapering, acuminate, the lateral chartaceous, glabrous above, minutely adpressed, pubescent beneath, lateral nerves 9-10 on either half, slightly arched tertiary sub-parallel, transverse, base rounded or sub-cordate. Racemes generally terminal and panicled. Flowers conspicuously bracteolate, calyx tube long, tuberculate, teeth long, the two upper connate Corolla bluish white. Stamens 2 adelphous. Ovary sessile, many ovuled, .Fruit pod.

Uses : Bark is used for making ropes and the thread is used for making fishing nets.

63. Millettia cinerea Benth. Fabaceae

Common Name : Rolang.

Distribution : Arunachal Pradesh.

A fairly large woody climber with a bushy habit, young branches pubescent, bark exuding a reddish gum. Leaves swollen at the base, stipules subulate, leaflets generally obovate oblong or elliptic pubescent beneath when young, soon becoming glabrescent, lateral nerves 7 - 9 on either half/more or less arched, petioles about 0.5 m long, stout. Racemes generally in terminal silky pubescent panicle. Stamens diadelphous. Ovary densely shaggy with white hairs. Style glabrous curled upward. Pod generally 1 - 2 seeded, but sometimes as many as 8 seeded.

Uses : Stem yields fibre for making ropes.

64. Naravelia zeylanica DC. Ranunculaceae

Common Name:Karopihoi, Jyrmailasam, Sangomgri, Chaglat.

Distribution : Throughout N.E. Region.

A woody climber with salcate, adpressed pubescent branches, leaves pinnately 2-3 foliate, the upper leaflets reduced into tendril. Leaflets usually two in unequal pairs, ovate rotundate, acuminate, rounded, entire or toothed. Inflorescence terminal or axillary panicle. Flowers greenish, stamens numerous. Fruit achene.

Uses : The stem is used for preparing ropes. The stem can be twisted into rough but strong ropes.

65. Oreocnide frutescens (Thunb.) Miq. Urticaceae

Common Name : Tillejuat, Thehoy - arong, Dieng - jei - thang sim. Thing - Khra - phang, Tappen.

Distribution : Throughout N.E. India particularly Assam and Meghalaya.

A small evergreen tree, young parts pubescent or tomentose. Bark greyish, warty with lenticels blaze red, exuding water, green fibrous. Leaves elliptic - oblong, acuminate usually entire, sometimes obscurely cernate, thinly chartaceous, sparsely puberulous above, pubescent beneath, lateral nerves 9 - 12 on either half, base obtuse. Stipule silky, hairy, deciduous. Flower heads dichotomously branched, cymose hispid. Male flower 3 - 4 merous stigma papillose, ciliate, the bracts become succulent and glossy when the fruit ripens.

Uses : Fibre is utilized for making fishing nets.

66. Oreocnide integrifolia (Gaudich.) Miq. Urticaceae

Common Name : Lipe, Kujjyet - Kee, Khilkila, Kilkra, Boree, Tane, Yesa.

Distribution : Eastern Himalayas, including Assam and adjoining states upto an altitude of 1500 m.

A small evergreen tree with pubescent or tomentose young parts. Leaves elliptic oblong, obvate - oblong acuminate entire or slightly crenate, stipules silky hairy, inter petioler. Flowers small, unisexual, dichotomously branched, cymose, hispid.

Uses : The bark is used for fishing nets, fishing lines and coarse clothing. The fibre resembles rhea (*Boehmeria nivea*) and therefore commonly called Ban rhea.

67. Polyalthia simiarum (Hook.f. & Thoms.) Annonaceae

Common Name : Bogakhamtau, Silemphang, Borsthi, Lansgaya.

Distribution : West Sikkim, Assam.

A large tree with light or dark grey bark. Leaves biparous, broad, thinly coriaceous, glabrous and shining above. Sometimes more or less pubescent along with midrib and nerves beneath, pinkish - brown and drooping when young, lateral nerves 12 - 18 on either half, arcuate, conspicuous beneath. Flowers yellowish green in fascicles from axils of fallen leaves or from tubercles on the branches, pedicles as long as the flowers, slender, minutely pubescent with a bract about the middle. Fruit orange red to blue black. Seeds ovoid, grooved and ribbed.

Uses : Stem bark yields fibre which is used for making ropes.

68. Pterygota alata Roxb. Sterculiaceae

Common Name : Boktha, Mi - bol.

Distribution : Throughout N.E. India.

A tall handsome, often buttressed tree up to 40 m in height and 3 m in girth with straight cylindrical hole up to 30 m. Bark greyish brown smooth. Leaves ovate cordate. Flowers brownish yellow in panicles, follicle sub - globose. Seeds many, oblong, compressed.

Uses : The wood is chiefly used for tea boxes. The bark yields a fibre used for cordages.

69. Sida cordifolia Linn. Malvaceae

Common Name : Berela, Bala, Favadbute, Kheranti.

Distribution : Throughout N.E. India.

A small undershrub, generally found near moist places. Leaves oblong, cordate, obtuse, petioles as long as blade. Peduncle jointed near the flowers.

Uses : The stem yields a fine white silky fibre much superior to jute and therefore, is of considerable potential. It is commercially valued.

70. Stephania glabra (Roxb.) Miers. Menispermaceae

Common Name: Barkulilahara, Kani-Kario, Lempok, Ganjaroo, Lempau.

Distribution : Sikkim, Nagaland.

A large climber with tuberous roots and rather fleshy branches, tubes globose, deep-yellow inside. Leaves orbicular, sometimes broader than long, usually peltate, often with repand or lobed margins, membranous, glabrous, basal nerves generally 9, intermediate nerves very slender and finely reticulate. Flowers yellow. Peduncle axillary. Fruit drupe globose.

Uses : It yields a fibre used for fishing - lines.

71. Sterculia villosa Roxb. Sterculiaceae

Common Name : Odul, Kanhlyem - Kung, Udhara, Olmak, Libak, Heppukbung, Tiengstar.

Distribution : Common in deciduous forest of Assam ascending up to 1000 m.

A large deciduous tree with spreading crown. Leaves crowded at the end of the branches. Flowers dull yellow in crowded drooping panicles. Males are more numerous than female. Calyx campanulate pubescent. Staminal column with 10 anthers. Gynophore 1/2 cm long, ovary globose, staminode in a ring beneath the ovary.

Uses : The inner layer of the bark yields a coarse and strong fibre which is used for cordages and for making rough ropes and bags. It is used and valued for breast bands of elephants for dragging timber.

72. Trema politoria (Planch.) Blume Ulmaceae

Common Name : Kuail, Amli, Chinch, Hitamg, Tetul.

Distribution : Throughout N.E. India.

A shrub or small tree, bark pale brown, smooth, dotted with lenticels, branchlets scabrid. Leaves oblong ovate - oblong or ovate - lanceolate, acute or acuminate, crenate, serrulate, membranous, scabrid and thinly hispid on both surfaces, base rounded, truncate or sub - cordate. Inflorescence cymes. Fruit drupe.

Uses : Bark is used for making ropes for all purposes. Leaves are very harsh to touch and are used as a substitute for sand-paper to polish wood, horn and ivory.

73. Urena lobata Linn. Malvaceae

Common Name : Kuray pat, Samthe, Sohbyrthit, Agro.

Distribution : Throughout N.E. India.

An undershrub, 1.5 m in height, stellate, pubscent. Leaves variable ovate to broadly cordate. Flowers pink with darker centre, one or more on axillary peduncles.

Uses : Stem gives a strong white fibre suitable for twine.

74. Urtica parviflora Roxb. Urticaceae

Common Name : Kanyali, Loure, Shokam.

Distribution : Throughout N.E. India.

A slender herb with angled branches. Leaves variable, ovate, cordate acuminate, crenate or serrate with stiff stinging hairs. Petioles long. Flowers monoecious in short peduncled cymes forming an erect terminal panicle. Males and fruiting sepals hispid.

Uses : The stem gives fibre which is used for ropes. It is said to be superior to that of jute in tensile strength and length of staple.

4

SPICES AND CONDIMENTS

75. Allium hookeri Thw. Liliaceae

Common Name : Yenum - napakpi.

Distribution : Throughout the N.E. Region.

A scapigerous herb, stem base clothed with long narrow membranous sheaths. Leaves globose, laxly many flowered. Flowers white. Sepals linear acuminate.

Uses : It is cultivated by the inhabitants and used as spice like onion or garlic leaves.

76. Allium tuberosum Rott. ex Spreng. Liliaceae

Common Name : Yenum - nakupi, Jyllang.

Distribution : Throughout the N.E. Region.

A scapigerous herb, elongate cylindrical with white fleshy root fibres. Leaves erect, narrow, linear flat, compressed scape 30 - 40 cm. Heads consisting of 20 - 40. Flowers white or pink.

Uses : It is cultivated by the inhabitants and used as a spice and as substitute for onion and garlic.

77. Alpinia galanga (L.) Swartz Zingiberaceae

Common Name : Vacha, Hulanger, Kanglu.

Distribution : Throughout N.E. Region.

A perennial herb with slightly aromatic tuberous rootstock. Leaves oblong-lanceolate, glabrous beneath. Flower small. Corolla segments linear - oblong, greenish white, clawed, obovate, emarginate, white black veined.

Uses : Cultivated by local inhabitants and used as spice.

78. Cinnamomum bejolghota (Buch.-Ham.) Sweet Lauraceae

Common Name : Dieng-la-si-sirmot, Pati-hunda. Seerang asing.

Distribution : Throughout the N.E. Region.

A large tree. Bark grey or brownish white, rough, up to 1.8 cm thick, blaze aromatic, yellowish or pale brown, turning darker brown on exposure. Leaves elliptic - oblong, obtuse, acute or acuminate glabrous, base 3 nerved, nervules, panicles large,long peduncled, subterminal, usually exceeding the leaves minutely pubescent or puberulous. Perianth lobes silky on both surface. Stamens and ovary sharply pubescent.

Uses : Used as spice.

79. Cinnamomum tamala Nees & Eberm. Lauraceae

Common Name : Huara, Tejpata, Deind-la-tyrpat, Tej-pat, Mensing.

Distribution : Throughout the N.E. Region.

A medium sized branching tree. Bark rather rough, dark grey, almost black or dark brown, blaze aromatic, reddish brown, darkening on exposure. Leaves alternate, sub - opposite ovate - oblong or elliptic to oblong - lanceolate, acuminate with the tip, often sub-falcate or acute glabrous pink when young, perianth silky, segments sub-equal up to 5 cm long, oblong or elliptic - oblong, silky pubescent on both surfaces. Stamens villous, ovary free, style glabrous. Fruit drupe.

Uses : Leaves are used as spice.

80. Curcuma amada Roxb. Zingiberaceae

Common Name : Amba haldi, Sigmambi.

Distribution : Throughout the N.E. Region.

A perennial herb, rootstock large, ovoid, sessile rhizome thick, cylindric, pale yellow inside, odour is similar to mango. Leaves large oblong lanceolate. Flowers pale yellow.

Uses : Cultivated for use as spice as ginger and also for medicinal uses. Fresh root is used as an ingredient of chutney. Tubers have an agreeable fragrant smell and aromatic taste. Infusion of the root is employed to give flavour of mango artificially to confectionery. The rhizome of *Curcuma amada* also yields 1.1% essential oil containing d - 1 pinene, ocimene 47.2%, sofrole 9.3 and unidentified substance.

81. Curcuma longa Linn Zingiberaceae

Common Name : Haldi, Turmeric, Kanreludunpa.

Distribution : Throughout the N.E. Region.

A perennial herb, rootstock large ovoid, sessile rhizome thick. Leaves sub - sessile. Flowers orange white.

Uses : Rhizomes are boiled, dried and made into powder which gives a yellow colour and is employed largely as a colouring agent and in pickles and curry powders. Good turmeric should be reddish orange in colour.

82. Eryngium foetidum L. Umbelliferae

Common Name : Awaphadigo, Dahania - khlan, Shannshai.

Distribution : Throughout the N.E. Region.

An aromatic glabrous erect perennial herb. Leaves spathulate, spinous, toothed. Flowers in heads and white. Fruit ellipsoid.

Uses : The plant is used as a substitute for coriander for flavouring rice and curries. It is very common in the Valley of Manipur growing in waste dry places and is also cultivated in kitchen garden for use as spice as well as medicine.

83. Hedychium coronarium Koenig Zingiberaceae

Common Name : Rubri.

Distribution : Throughout N.E. Region.

A stout rhizomatous herb. Leaves lanceolate, large, variable in breadth. Flowers white or yellow, fragrant imbricately terminal capitate spikes.

Uses : It is cultivated in gardens as a pot herb for the use as spice as well as medicine. The flower yield 07% of a dark maroon brown wax like concentrate with a strong spicy odour. It gives 50.0 - 57.8% of a thick oily substance of dark orange brown colour and hard brownish black wax.

84. Hedychium spicatum Buch. - Ham. ex Smith
Zingiberaceae

Common Name : Ban Haldi, Kapur - Kachri, Sedhu, Viteize.

Distribution : Throughout the N.E. Region.

A perennial rhizomatous herb. Leaves oblong or oblong lanceolate up to 30 cm long, very variable in breadth, glabrous. Spike dense flowered, bracts large oblong, calyx shorter than the bracts. Flowers white.

Uses : Used as spice and grown as garden herb.

85. Illicium griffithii HK. f. & Thoms. Illiciaceae

Common Name : Star-arise.

Distribution : Up to a height of 1500 - 2000 m, chiefly in gorges and ravines.

A large aromatic evergreen shrub 4 - 6 m high, branches somewhat angular, glabrous, shining. Leaves ovate, elliptic or lanceolate, shortly acuminate acute at base, coriaceous, glabrous, shining above, dull brown beneath. Flowers axillary or terminal. Sepals 4 - 6 orbicular acute. Petal 15 - 20 orbicular not pointed, fleshy stamens 20 - 24, erect, filaments very short, anthers broad blunt, carpels about 20 in a single whorl, fruit follicles.

Uses : The fruit is aromatic and is used as spice. It has bitter and astringent taste and is reported to be poisonous. It is stimulant and carminative. It yields an essential oil on steam- distillation.

86. Murraya koenigii (L.) Spreng. Rutaceae

Common Name : Curry leaftree, Mitha neem, Karhi neem, Kairsingi.

Distribution : From Garhwal to Sikkim and N.E. Region up to a height of 1500 m.

A handsome, aromatic more or less, deciduous shrub or small tree up to 6 m in height. Bark dark brown or almost black. Leaves imparipinnate. Leaflets 9-25, ovate lanceolate or somewhat rhomboid irregular crenate-dentate acuminate, obtuse or acute, almost glabrous above, pubescent beneath, gland dotted, strongly aromatic. Flowers in terminal corymbose cymes white fragrant, berries purplish black when ripe.

Uses : Leaves are used for flavouring curries and condiments and as natural flavouring agent in south Indian food i.e. sambhar, rasam and curries.

87. Zanthoxylum armatum DC. Rutaceae

Common Name : Tumree, Nepati, Dhania, Jaiur, Nangry.

Distribution : Throughout the N.E. Region.

Tree up to 8 m tall, prickles straight woody on old stem, young twigs glabrous. Leaflets 5 - 11 elliptic to oblong, lanceolate, cuneate at base, acute to bluntly acuminate serrulate, lateral nerves 10 - 16 pair, rachis 6 - 15 cm, winged. Flowers pale yellow, follicles 1 - 2 free sub compressed.

Uses : It is a spice which is substitute for coriander. Various parts of the plant are used in India and China for seasoning food. The leaves are eaten in fritters. The fragrant twigs are chewed for cleaning the mouth of bad breath.

5

ESSENTIAL OIL PRODUCING PLANTS

88. Acorus calamus Linn. Araceae

Common Name : Bach, Themepru.

Distribution : Throughout the N.E. Region.

A robust water loving perennial herb up to 2 m high. Rhizome creeping, branched, containing aromatic cells. Leaves linear, grass green, parallel veined, spadix 5 - 10 cm, slightly curved.

Uses : The dried rhizome yields 1.5 - 2.7% natural yellow, aromatic essential oil Ascerone having an agreeable odour. It also contains a glucoside Acronin. The fresh aerial parts yield 0.1% volatile oil, the unpeeled roots however, give a much better yield from 1.5 - 3.5%.

89. Alstonia scholaris R. Brown Apocynaceae

Common Name : Sing - gar - asing, Sotiana, Buchong, Chhatian, Ealilam.

Distribution : Throughout the N.E. Region.

A tall evergreen tree, branches whorled, trunk buttressed or fluted. Bark grey with large horizontal lenticel - dark and light vertical fissures. Leaves obovate oblanceolate or elliptic - oblong, emarginate, rounded or bluntly acùminate, coriaceous, glabrous shining above, pale beneath, lateral nerves. Flowers greenish - white in compact umbellately branched, pubescent, panicled cymes. Fruit follicles.

Uses : Flowers yield an essential oil.

90. Aphanamixis polystachya (Wallich) Parker Meliaceae

Common Name : Akawatong, Harinhara, Tiktaraj.

Distribution : Throughout N.E. Region.

Trees 5-10 m with spreading crown, Leaves crowded at ends of branchlets, 30-70 cm, leaflets 9-19, oblong obovate or oblanceolate, obliquely acute at base, caudate, lateral nerve 12-15 pairs, panicles as long as leaves. Flowers dull white, sessile. Capsules obovoid, seeds black covered with scarlet red aril.

Uses : Yields an essential oil and also semidrying oil. The oil is used as liniment in rheumatism.

91. Aquilaria agallocha Roxb. Thymelaeaceae

Common Name : Sasi, Agaru, Agor.

Distribution : Throughout N.E. Region.

A large evergreen tree, young shoots silky, leaves oblong lanceolate, elliptic to obovate - lanceolate, caudate - acuminate, glabrous or nearly so, base acute, lateral nerves numerous and joint, petiole long. Flowers white, bisexual in terminal umbel or cymes which are sub - sessile or shortly peduncled, inflorescence silky perianth persistent in fruit, campanulate, densely villous within, lobes 5, spreading about as long as the tube, rounded, scales above the stamens 5, densely villous connate at the base. Stamens 10, subsessile. Ovary also subsessile, villous, 2 celled Stigma large, subsessile. Fruit capsule.

Uses : The wood of the plant yields an essential oil called *chuwah* or *Agar Attar* used in perfumery. True agar is hard brown and rich in resin. It is prized mainly as incense by the Parsees and Arabs.

92. Butea minor Buch.-Ham. ex Baker Fabaceae

Common Name : Dieng - la - palang.

Distribution : Throughout N.E. Region.

An erect shrub with long switchy branches covered with grey -silky tomentum. Common petiole, 10-20 cm long, channelled, tomentose, stipules about 2.5 cm broad, semicircular, often persistent. Leaflets often up to 30-45 cm, glabrous above, bright grey-silky beneath, lateral nerves 7-9 on either half, very oblique. Inflorescence racemes.

Uses : It provides an oil which can be used for manufacture of paints and after hydrogenation high quality soaps.

93. Cinnamomum camphora (L.) J. Presl. Lauraceae

Common Name : Kapur, Karpur.

Distribution : Throughout N.E. Region.

Small tree with spreading crown. Bark black. Leaves alternate, ovate, elliptic, petiole acuminate, glabrous shining above, white or glaucous beneath. Flowers yellow axillary panicles, perianth lobes - deciduous, stamen 9, ovary sessile.

Uses : The plant is a source of camphor which is distilled from the chips of old wood. In Sri Lanka the camphor is distilled from young shoots. Camphor is extensively employed in external applications in the treatment of muscular strains, rheumatic conditions and inflammation. In conjunction with menthol or phenol it relieves itching of the skin.

94. Cinnamomum glaucescens (Nees) Meissn. Lauraceae

Common Name : Suriarong, Kanhuroi, Gondroi phang.

Distribution : Throughout N.E. Region.

A large to middle sized trees, with dark brown or grey furrowed bark, usually brown with darker streaks. Buds scales broadly ovate or sub - orbicular, imbricate, brown silky tomentose, inflorescence and juvenile shoots brown, tomentose, older branchlets glabrous, green. Panicles usually crowded at the base of the current years shoots. Flowers densely tomentose.

Uses : It yields an essential oil. The wood gives 2.95% of a clear pale yellow essential oil similar to the odour of sassafras (*Sassafra officinale / Laurus sassafras*).

95. Curcuma aromatica Salisb. Zingiberaceae

Common Name : Base, Banhardi, Jalam olike.

Distribution : Abundant at lower altitudes throughout N.E. Region.

Known as "wild turmeric" it is a herb with light yellow (internally orange - red) rhizomes having camphoraceous odour. Rhizomes of this plant are sometime used as a substitute for turmeric (*Curcuma longa*) but not as a condiment. They are externally applied in combination with astringents, bitters and aromation to bruises and sprains.

Uses : On steam distillation, the rhizome yields an essential oil (about 6.0%) which is greenish brown in colour and has a camphoraceous odour. The oil contains sesquiterpenes (65.5%) two monocyclic tertiary, sesquiterpene alcohols (22.0%), d - camphene (8%), the colouring matter has been identified as curcumin.

96. Gaultheria fragrantissima Wallich Ericaceae

Common Name : Jalan - thrait.

Distribution : Meghalaya 1500 - 2000m.

An evergreen shrub up to 4 m high, leaves faintly aromatic and smelling of methyl salicylate (oil of winter green) when bruised, oblong - lanceolate to elliptic lanceolate, acute or acuminate, serrate, coriaceous, glabrous and dark green above, pale with black dots below. Flowers white in axillary racemes. Calyx white, stamens 10 attached to the base of the corolla, ovary 5 celled, ovules many in each cell.

Uses : It produces oil similar to that obtained from Canadian winter green, which is distilled from the leaves. This oil is reported to be superior to the Canadian variety. Oil of Gaultheria is stimulant, carminative and antiseptic. It is applied externally in the form of liniment or ointment in rheumatism, sciatica and neuralgia.

97. Hedychium spicatum J.E. Smith Zingiberaceae

Common Name : Ban Haldi, Kapur Kachari.

Distribution : Throughout N.E. Region.

A perennial herb. Leaves oblong or oblong lanceolate up to 30 cm long, very variable in breadth, glabrous. Spike dense flowered, bracts large. oblong, Calyx shorter than the bracts. Flowers white yellow.

Uses : The rhizome on steam distillation yields an essential oil (4%) with a characteristic odour. The principal constituent is the ethyl ester of p - methoxy cinnamic acid (67.8%). The other constituents are ethyl cinnamate (10.2%) and uneole (6.0%). The oil may be used as perfume for soap, hairoil, facepowder and incense. Besides the essential oil, the rhizomes contain starch (52%), organic acid including resinic acid and a glycoside and ash (4.6%). The dried fruit is reported to be added to soften meat and pulse during cooking.

98. Impatiens amplexicaule Edgew. Balsaminaceae

Common Name : "Chi", Phykutli.

Distribution : N.E. Region particularly Arunachal Pradesh.

A tall branched tree. Leaves sessile, lower opposite oblong - lanceolate, upper alternate ovate lanceolate, amplexicaul crenate, serrate. Stipules pubescent. Flower umbel or raceme standard orbicular 2 - lobed wings 2 - lobed, saccate acute or obtuse, spur short inflexed, capsule slender horizontal.

Uses : Prospect of commercial exploitation of the oil as varnish is indicated. As the fruits of the plants are sensitive to touch and a large quantity of seed is wasted, a planned collection of seed is recommended before dehiscence of the capsule.

99. Juglans regia Linn. Juglandaceae

Common Name : Akhrot, Akhor, Okhar, Taga.

Distribution : Throughout N.E. Region.

A large aromatic deciduous tree, shoot velvety, bark grey, longitudinally fissured. Leaves imparipinnate long, thickly tomentose when young. Leaflets subsessile opposite or sub - opposite elliptic oblong, often oblique usually entire, acute or acuminate, coriaceous, pubescent along the nerves beneath or with tuft of hairs at the axils of the nerves, lateral nerves, 15 - 20 on either half, base rounded or obtuse. Male catkins green often in pair, Female flower 1 - 3. Calyx obscurely 4 toothed. Petals 4, green stigma more or less fleshy. Fruit ovoid, glabrous or pubescent, green with yellow dots, pericarp leathery aromatic.

Uses : It yields an essential oil.

100. Juniperus recurva Buch.-Ham.ex D. Don Cupressaceae

Common Name : Sukkpa, Padmak.

Distribution : Throughout N.E. Region.

A small or middle sized tree. Leaves dimorphous, subulate and purient on young shoots scale like or mature branches. Large resinous glands on the back. Flowers monoecious. Male catkins at the end of the branchlets, Fruit almost globose, blue - black when ripe very resinous.

Uses : Wood of *Juniperus recurva* is one of the best material among Indian timbers for making pencils. The essential oil is derived from the fruits. It is a good substitute for imported Juniper oil.

101. Lindera neesiana (Wallich ex Nees) Kurz Lauraceae

Common Name : Jamkañuesing.

Distribution : Throughout N.E. Region.

A middle sized tree, aromatic glabrous. Leaves ovate oblong, lanceolate, rarely sub - orbicular usually acuminate, entire membranous, glabrous, lateral nerves 4 - 6 on either half, slender, angular, base obtuse, rounded or cuneate, slightly oblique, irregularly 3 nerved. Head 5 - 6 flowered, solitary or clustered globose. Perianth segments almost orbicular, membranous, almost glabrous, stamens 9, filament biglandular, fruit globose.

Uses : Essential oil is obtained from the flower.

102. Litsea cubeba (Lour.) Pers. Lauraceae

Common Name : Mejankei, Dieng-si-ing, Zeng-jil, Akmer, Soh-slyang.

Distribution : Throughout N.E. Region.

A small tree, deciduous, aromatic. Leaves lanceolate or ovate - lanceolate, glabrous above, glaucous beneath. Flowers in capitate umbels, solitary or in corymbs. Perianth segments 6, membranous. Fruits small globose.

Uses : The leaves of the plant are steam distilled and the oil is extracted. Fruits are aromatic and yield on distillation a volatile oil with an odour reminiscent of verbina oil. The oil has recently gained importance in China as a source of citral and is reported to be particularly suitable for the production of methyl ionone. Citral and methyl ionone derived from the oil possess a finer odour than those from lemon grass oil.

103. Madhuca longifolia (Koen.) Macb. Sapotaceae

Common Name : Mahua, Garang, Irum, Kayonasoso.

Distribution : Throughout N.E. Region.

A middle sized tree. Leaves clustered towards the end of the branchlets, elliptic - oblong or obovate, oblong, glabrous, chartaceous, lateral nerves 23 - 25 on either half, sub - parallel, slightly arcuate at the margin, base narrowed into the petiole, unequal, oblique. Ovary glabrous. Style glabrous, long exerted. Fruit a berry.

Uses : Fruits give an oil.

104. Mesua ferrea Linn Guttiferae

Common Name : Mangitong, Micharne, Nahaur - araung.

Distribution : Throughout the N.E. Region.

An evergreen middle sized handsome tree with dense conical crown. Leaves opposite decussate, narrowly lanceolate or lanceolate oblong, acuminate or cuspidate white or glaucous beneath or glabrous red when young, Flowers solitary or paired, Sepals 4 in two pairs, Petals 4 white, Stamens many forming a globose yellow mass in centre of the flower.

Uses : The oil obtained from the fruit and seed is used in the manufacture of soap and other items.

105. Nardostachys jatamansi (D. Don) DC Valerianaceae

Common Name : Jatamansi, Mansi.

Distribution : 4000 - 5000 m. (Alpine Himalayas).

A perennial herb of economic importance known as "spikenard" or "Indian Nard". The plant is an erect perennial herb, 10 - 60 cm high with white heads occurring in moist alpine meadows. Root of the plant contains an essential oil which has a fine flavour similar to patchouli and valerias oils.

Uses : A valuable essential oil (Oleum jatamansi) is isolated from rhizome. The fresh root is also used in the preparation of medicinal oils and perfumery.

106. Perilla frutescens (L.) Britton Labiatae

Common Name : Arim, Angami.

Distribution : Throughout the N.E. Region.

A strongly aromatic bush, stem, leaves and petiole covered with grey hair. Leaves ovate or sub - orbicular, coarsely serrate acuminate, membranous, scabrous above, hairy beneath, specially along the nerves, petiole long. Flowers small white in axillary or terminal spicate racemes. Calyx villous. Corolla campanulate. Stamens 4 sub - equal erect.

Uses : Oil obtained from the flower is used in the manufacture of the soap. Its leaves are used in Japan for colouring plum preparations. They contain an anthocyanin, Perrillanin chloride. The flower buds are sweet and eaten raw or cooked as vegetables.

107. Ricinus communis Linn. Euphorbiaceae

Common Name : Inkiang, Sah - rasam, Arandi, Gab, Lulucha.

Distribution : Throughout the N.E. Region.

An evergreen soft - wooded shrub, young part covered with a glaucous bloom. Leaves alternate, 1 - 4 cm across, palmately lobed, peltate serrate, membranous, petiole stout, hollow, as long as leaves. Flowers monoecious. Stamens numerous. Ovary 3 celled. Style 3, plumose 2 fid. Capsule globose, echinate. Seeds with large caruncle, testa crustaceous.

Uses : The seeds yield the well known castor oil.

108. Xeromphis spinosa (Thunb.) Keay Rubiaceae

Common Name : Mon, Ruhiaong, Mawnakata, Thelkha.

Distribution : Throughout the N.E. Region.

A large deciduous shrub or sometime a small tree with stout axillary spines. Bark dark brown or grey, somewhat rough. Leaves generally fascicled, obovate or oblanceolate, roundish or obtusely short, acuminate or acute, pubescent, rough or glabrous, lateral nerves 5 - 7 on either half, sub - parallel, base narrowed into a short marginate petiole. Flower sub - sessile greenish white or yellowish, fragrant, solitary or 2 - 3 together on short lateral branchlets calyx strigose, tube long, campanulate, segments variable ovate or lanceolate or spathulate foliaceous as long as the tube. Corolla covered with adpressed silky hairs, tube not exceeding the calyx. Berry soft.

Uses : Flowers yield an essential oil with fragrance reminiscent of gardenia oil but the perfume is not commercially prepared. The roots possess marked insecticidal and insect repellent properties.

6

ETHNOTOXIC PLANTS

109. Acacia pennata (L.) Willd. Mimosaceae

Common Name : Sirengkhi, Ther-chi-kli, Jermai-sheih- lyng-kshaih.

Distribution : Throughout the N.E. Region.

A large prickly climber, bark dark brown to grey, often with horizontal cracks, blaze pink to red, fibrous, branches with about 5 rows of small more or less recurved prickles. Young shoots, inflorescence, leaf rachis and pinnae pubescent. Leaf rachis 20 cm long with a large cup - like gland near the base and 2 - 4 smaller glands between the upper pinnae, usually prickly. Flowers pale white in pedunculate heads on large terminal leafy panicles. Calyx glabrous corolla minute slightly exceeding the calyx.

Uses : The fruits and stems are used as fish poison. The bark contains 9% of tannin and is used for tanning, produces stiff leather of some what harsh structure and reddish brown colour.

110. Albizzia procera (Roxb.) Benth. Mimosaceae

Common Name : Sidak, Korai.

Distribution : Throughout the N.E. Region.

A tall handsome sub - deciduous tree, attaining a height of 20 - 25 m and is about 2 - 2.5 m in girth, bark nearly smooth, yellowish, grey or brown outside, peeling off in thin irregular plates. Leaf rachis long, basal gland large, oblong or oval. Pinnae 2 - 6 pairs, leaflets 6 - 16 pairs, oblique oblong - ovate or rhomboid oblong, obtuse, silvery when young, both surfaces thinly adpressed pubescent or glabrescent with age, under surface paler, thinly coriaceous, midrib diagonal near the lower edge except at the apex. Flowers sessile, yellowishwhite, calyx tubular, glabrous outside teeth unequal, triangular, Corolla lobes oblong - lanceolate pubescent outside. Stamens filaments yellowish, staminal tube about as long as corolla tube.

Uses : The bark of the plant is used as fish poison.

111. Anamirta cocculus Wight & Anr. Menispermaceae

Syn. *Anamirta paniculata* Colebr.

Common Name : Kakamari.

Distribution : Throughout the N.E. Region.

A powerful climber, wholly glabrous, often sending down numerous rootlets. Leaves ovate shallowy cordate or truncate, coriaceous, glabrous. Flowers green scented. Male apetalous but with deciduous sepals, stamens numerous with 2 celled anthers. Female flowers with club shaped staminodes carpels usually 3 on short gynophore.

Uses : The fruit is very bitter and is used as fish poison. The oil of the seed is useful for industrial purposes. An ointment prepared from berries is used as an insecticide. The dried ripe fruits produce the drug Cocolus fructus also called Cocculus indicus.

112. Barringtonia acutangula (L.) Gaertn. Barringtoniaceae

Common Name : Hinyot, Paniamara Bheh, Chimchong - arong.

Distribution : Throughout the N.E. Region ascending to 1000 m usually on banks of streams, edges of swamp and moist places.

A small to medium sized tree. Leaves alternate crowded at the end of the branches, obovate or oblanceolate apex rounded or sub - acute, coriaceous, glabrous.Flowers deep red fragrant in slender, drooping, glabrous racemos 3 - 7 cm long calyx tube funnel shaped, petals 4 adnate to the base of the staminal tube. Stamens in several series, red, connate below, ovary inferior 2 celled.

Uses : The tender leaves are edible. The bark, root and seeds are reported to be used as fish poison. They probably contain saponin.

113. Buddleja asiatica Lour. Buddlejaceae

Common Name : Agia - chita.

Distribution : Throughout the N.E. Region, ascending upto a height of 1500 - 2000 m.

Shrub, Leaves simple narrow lanceolate acute, dark green above, white tomentose beneath. Flowers small white in panicles or cymes. Calyx campanulate, Corolla arcuate, anther sub - sessile stigma clovate.

Uses : Leaf and shoots are used as fish poison.

114. Canthium dicoccum (Gaertn.) Teysmann & Binnedijk
Rubiaceae

Common Name : Bathling, Ramnisethar.

Distribution : Assam, Meghalaya, Nagaland and Manipur.

A stout large evergreen shrub with spreading or drooping branchlets. Leaves entire, elliptic or ovate lanceolate abruptly acuminate, coriaceous quite glabrous, glossy. Flowers pentamerous white fragrant in dense axillary or umbellate cymes sessile or shortly peduncled. Calyx truncate. Corolla tube companulate. Stamens on mouth of corolla, Style long. Fruit globose blackish or purple.

Uses : Bark is used as fish poison.

115. Cassia mimosoides Linn. Caesalpiniaceae

Common Name : Sikkum.

Distribution : Throughout the N.E. Region.

Usually a diffuse, sometimes an erect undershrub, often with spreading branches, leaf rachis and inflorescence clothed with adpressed hairs. Leaves slender with a sessile gland close to the lower pair of the leaflets, stipules subulate with a broad base. Leaflets crowded and overlapping, linear oblong, acute, obliquely mucronate, rigid, nerves somewhat prominent beneath, often with a few hairs near the margin on the under surface. Flowers solitary or in pairs, pedicels axillary or extra axillary, bracts one pair at the base of the pedicels, resembling the stipules, bracteoles 2, a little below the caylx about half the size of the bracts. Caylx segments unequal, long, ovate, acuminate, thinly pilose, outside. Petals long, yellow elliptic obtuse. Stamens usually 10 alternately shorter and long, strap shaped with scattered short hairs outside, septas between the seeds, compressed, obliquely oblong.

Uses : The crushed leaves and stem are used as fish poison.

116. Croton caudatus Geisel Euphorbiaceae

Common Name : Lota - mahudi, Ghahe - lewa, Kum - kum - arong.

Distribution : Throughout the N.E. Region.

A large scandent shrub, young parts stellately bristly hairy. Bark brownish black old bark exfoliating in pieces exposing a whitish surface, blaze greenish white, wood whitish. Leaves ovate, sub -orbicular or ovate acuminate coarsely toothed or crenate - serrate, sub - coriaceous, scabrous above, stellately pubescent beneath, hairy on both surfaces, basal nerves 3 - 5, base rounded, cordate or almost truncate, basal glands present. Flowers in fascicles. Male flowers tomentose, petals as long as sepals, stamens 18 - 30. Female flowers : sepal ovate, oblong, petals very small, ovary stellately woolly, style 2, capsule obscurely trigonous scabridly rusty - pubescent.

Uses : Seeds of the plants are used as arrow poison.

117. Croton tiglium Linn. Euphorbiaceae

Common Name : Koribih, Dieng - kymbat - lasam.

Distribution : Assam, Meghalaya upto a height of 1000-1500 m.

A small evergreen tree, young leaves brown. Leaves elliptic-oblong, acute or acuminate more or less serrate. Male flowers sepals bearded at apex. Petals woolly, stamens 15 - 25. Female flowers : sepals sparsely hairy at base within ovary stellately hispid.

Uses : Seeds are used as fish poison. The endosperm is reddish brown and is rich in oil. The oil which amounts to 30 - 45% of the whole seed is obtained commercially by extraction.

118. Derris elliptica (Wallich) Benth. Fabaceae

Common Name : Bakol - biti, Makoi - sopa, Etam - chali.

Distribution : Throughout the N.E. Region.

A long woody climber with long velvety branches. Leaflets oblong, obovate - oblong often blunty short acuminate glabrous above softly pubescent beneath. Racemes terminal or from upper axis, lax flowered. Rachis and branches brown, pubescent. Flowers red or whitish.

Uses : The root is used as fish poison and bark is chewed with betel leaves. The roots contain large percentage of rotenone which is known to be valuable insecticide. There is therefore demand for the roots for agricultural insecticides.

119. Derris scandens Benth. Fabaceae

Common Name : Gonj, Noalota.

Distribution : Throughout the N.E. Region.

A large woody climber with lenticellate branches, young parts rusty pubescent. Leaves imparipinnate oblong or obovate - oblong, apex often tapering, obtuse sub - coriaceous, glabrous above, puberulous beneath, secondary nerve somewhat obscure, 6 - 8 on either half, Flowers white or pink in fasciles on long axillary and terminal rusty pubescent racemes. Calyx long truncate, silky pubescent. Corolla exerted sub -orbicular slightly cordate near the claw.

Uses : Root is used as a fish poison.

120. Edgeworthia tomentosa Nakai Thymelaeaceae

(Syn.) *Edgeworthia gardneri* (Wallich) Meissn.

Common Name : Ruteng.

Distribution : Throughout the N.E. Region.

A large woody shrub, branches long. Leaves oblong elliptic lanceolate, acute acuminate glabrous above somewhat pilose beneath, base acute. Flowers yellow in dense hemispherical heads, perianth sub - villose, tomentose, lobes ovate, obtuse, stamens 8, biseriate raceptacle tubercular silky a villose. Fruit included in the base of the perianth, pericarp coriaceous, testa crustaceous.

Uses : The fruit of the plant is used as fish poison.

121. Engelhardtia polystachya Radlk. Juglandaceae

Common Name : Phekphe, Mowa, Sarong - phang, Lewa.

Distribution : Throughout the N.E. Region.

A middle sized or large tree, stem fluted, bark dark brown. Leaves usually paripinnate, red when young, leaflets opposite, elliptic or oblong, acute or obtuse, coriaceous, glabrous above clothed with golden peltate glands beneath. Male flowers in slender terminal spikes, often fasciculate bracts membranous.

Uses : Bark is used as fish poison.

122. Eupatorium odoratum Linn. Asteraceae

Common Name : Telimbabo.

Distribution : Throughout the N.E. Region.

Much branched shrub, 1-2 m internodes upto 10 cms. Leaves deltoid or rhomboid-ovate, entire and abruptly acute at base, long acuminate, sub-entire at margins, 20-25 flowers, in trichotomous convex corymbs. Involucular bracts many - serrate, pale or stramineous, Corolla pale blue to white.

Uses : Whole plant is used as fish poison

123. Ficus elastica Roxb. ex Hornem Moraceae

Common Name : Chang-krang, Deing-jari, Phrapramkhet, Dewak-arong.

Distribution : Throughout the N.E. Region.

A gigantic tree, quite glabrous, sending down numerous aerial roots, trunk strongly buttressed, usually epiphytic under natural condition. Leaves elliptic to oblong, entire bluntly and abruptly caudate, coriaceous, shining, lateral nerves very fine, numerous, almost parallel, base rounded or narrowed, obscurely 3-5 nerved, petiole long, stipules lanceolate, flaccid, reddish, almost half as long as the leaves. Receptacles enclosed in hooded involucres which fall leaving saucer shaped base under the fruit, sessile in pairs, smooth greenish yellow, ovoid - oblong. Male flowers pedicelled, sepals 4, ovate, anther sessile. Female flowers mostly sessile, sepals 4, ovary smooth, style sub - terminal, hooked, style long stigma sub - capitate, achenes ovoid.

Uses : Fruits are used as fish poison. The bark of the tree yields a caoutchouc like material of poor quality. The leaves contain caoutchouc.

124. Ficus silhetensis Miq. Moraceae

Common Name : Dieng - sormeblang, Bhuin - khannak.

Distribution : Throughout N.E. Region.

A shrub sometimes a long tree, rarely straggling, young parts pubescent. Bark ash grey, smooth, blaze greenish white. Leaves oblanceolate, elliptic - lanceolate or elliptic, entire minutely undulate, acuminate, membranous, glabrous, lateral nerves 4 - 8 on either half, slender stout, reticulation fine. base rounded, 3 nerved, petiole long hairy. Male flowers pedicellate, sepals 3, stamens 2, anther elongate, Female flowers almost sessile, sepals 3, style long, lateral curved. Receptacle axillary, solitary or in pairs, globose and reddish when ripe achene obovoid, flattened, papilose purplish, variegated.

Uses : Fruit extract is used to intoxicate the fishes.

125. Gynocardia odorata R. Br. Flacourtiaceae

Common Name : Chalmogra, Dieng-soh, Pharling, Sibe-tulpi.

Distribution : Most common throughout N.E. Region in evergreen forests.

Large evergreen tree with slender branches. Bark grey and warty outside. Leaves biparous oblong, coriaceous acuminate. Flowers dioecious, pale yellow in few flowered axillary fascicles or in large branches from the trunk. Calyx 5 lobed leathery, Petals fleshy, stamens about 100.

Uses : Fruit pulp is used for stupefying fishes.

126. Hydnocarpus kurzii Warb. Flacourtiaceae

Common Name : Chaulmogra, Dieng - soh - lap, Thibongthar.

Distribution : Throughout N.E. Region.

Evergreen tree, Leaves coriaceous, petiole swollen and slightly geniculate at the upper end. Flowers yellow in few flowered axillary cymes. Sepals 4 in 2 pairs. Petals 8 in 2 rows. Stamens 20 - 30, ovary in female and bisexual flowers villous. Fruit globose with velvety rind.

Uses : Fruits are used as fish poison.

127. Maesa indica (Roxb.) Wall. Myrsinaceae

Common Name : Baching, Puremo-kung, Samnakhatok.

Distribution : Throughout N.E. Region.

Shrubs or trees, leaves ovate - elliptic to oblong lanceolate, acute or cuneate at base, acute or acuminate, lateral nerves 6 -

12 pair, petioles 1 - 2 cm, Raceme axillary simple or branched, dense flowered. Fruits globose, cream yellow to dull brown.

Uses : Leaves of plant are used as fish poison.

128. Millettia auriculata Baker ex Brandis Fabaceae

Common Name : Hololokha, Bologlakha, Gurari, Gauj.

Distribution : Throughout N.E. Region.

Semi erect shrub with climbing branches and finely silky young parts tomentose. Leaves paripinnate, large, crowded at the ends of branches. Leaflets obovate oblong or elliptic acuminate, minutely pubescent beneath, Racemes simple, axillary. Flowers dull white.

Uses : Root is used as fish poison. The powdered root is also used to kill vermin in cattle. The root is reported to possess insect, repellent properties.

129. Millettia pachycarpa Benth. Fabaceae

Common Name : Holasi, Mak-khal, Ru - gogai, Arru, Kharia.

Distribution : Meghalaya upto 1500 m Tripura, Assam, Manipur.

Climber, large, brown velvety, Leaves deciduous, leaflets, terminal obovate, lateral oblong, lanceolate glabrous above brown tomentose beneath, cuneate at the base. Flowers in axillary racemes or terminal panicles. Calyx brown tomentose.

Uses : Root extract is used as fish poison.

130. Randia fasciculata DC. Rubiaceae

Common Name : Suho, Pulikant.

Distribution : Throughout N.E. Region.

A handsome spreading shrub or a small tree with short axillary spines, branchlets pubescent. Bark greenish - grey, warty thick having minute vertical lines. Leaves oblong - lanceolate, elliptic or ovate, acute or acuminate, entire, membranous, dark green and shining, glabrous, lateral nerves 9 - 11 on either half, usually rounded at the base, petiole long, stipules narrow lanceolate. Flowers white, fragrant, bracteate, usually in 3 flowered cymes at the ends of the branchlets, sometimes axillary and fasciculate, calyx pubescent, teeth linear deciduous, corolla tube slender, lobes broad, oblong often apiculate hairy at the base. Anthers sub - sessile at the mouth of corolla - tube, stigma slightly exerted, fusiform. Berry purplish black when ripe.

Uses : Root and fruits are used as fish poison.

131. Schima wallichii (DC.) Korth. Theaceae

Common Name : Noga-bhe, Dieng - shyrngan, Dieng nganbuit.

Distribution : Throughout N.E. Region.

Large tree with spreading crown. Bark brownish grey to blackish, with longitudinal fissures. Leaves ovate or oblong to lanceolate acute or acuminate, entire, thinly coriaceous, glabrous above puberulous beneath. Flowers white, scented axillary, pentamerous, Fruit a loculicidal capsule, globose, supported by persistant calyx.

Uses : Bark is used as fish poison.

132. Tephrosia candida (Roxb.) DC. Fabaceae

Common Name : Bilookhoni, Bangali-dadigdiga, Samander, Bolmendu, Dieng-toh.

Distribution : Entire N.E. Region ascending upto 1200 m or more often grown in tea garden as a soil ameliorant and also hedges.

Shrub 2 - 3 m high, bark brown, Leaves imparipinnate oblong or obovate - oblong, acute at the base mucronate at apex sub - coriaceous, glabrescent above, pubescent beneath. Racemes terminal and axillary. Flowers creamy white, standard obovate densely silky outside. Stamens diadelphous.

Uses : Bark is used as fish poison.

133. Trevisia palmata (Roxb.) Vis. Araliaceae

Common Name : Tago, Dieng-la-kor, Chenathong.

Distribution : Throughout N.E. Region.

Tree or large shrub. Leaves orbicular deeply palmatifid or palmatisect into 3 - 9 lobes, truncate or sub cordate at base, lobes acuminate serrate or irregularly lobed. Main nerves 3 - 9, palmate, petioles sheathing at base panicles terminal. Fruits sub - globose.

Uses : Fruits are used as fish poison.

134. Uraria crinita (L) Desr. Fabaceae

Common Name : Diong - kha - riu.

Distribution : Throughout N.E. Region.

An under shrub, Leaves trifoliate, leaflets oblong, acute, rounded at base coriaceous, glabrous above, pubescent along the nerves below. Racemes terminal, simple shaggy with spreading brown bristles. Corolla purple.

Uses : Root is used as fish poison.

135. Viburnum corylifolium HK. f. & Thomas Sambucaceae

Common Name : So - lang - ksew, Sohlang.

Distribution : Throughout N.E. Region.

A straggling shrub upto 4 m in height. Branchlets and inflorescence hirsute. Leaves elliptic, acute, dentate, membranous, underside densely clothed with soft hairs, sub cordate or rounded, lateral nerves 5-8 cm on either half, sub-parallel, Flowers sub-orbicular sessile, calyx stellate hairy outside, segments obscure, Corolla, white, rotate, hairy outside, glaucous within, 5 lobed, lobes sub - acute or orbicular. Stamens exserted anther densifixed Ovary 1 celled, style narrowed from a broad base. Stigma 3 lobed, drupe globular, bright red, Seed flat, shallowly grooved, albumen uniform.

Uses : Fruits are used as fish poison.

136. Viburnum foetidum Wall. Sambucaceae

Common Name : So - lang - ksew, Dieng-Soh-beng.

Distribution : Throughout N.E. Region.

A shrub upto 3 - 4 m high. Bark greyish, branchlets petioles and inflorescence clothed with stellate hairs. Leaves rhomboidal or elliptic lanceolate, mucronate, coriaceous, glabrascent above, pubescent on nerves beneath, lateral nerves 3 - 4 on either half, lowest pair basal, base cuneate, petiole long, Flowers small, in compound pedunculate umbels, bracts numerous, Calyx tube glabrous, teeth minute, triangular, Corolla short, white, Drupe scarlet red. Seeds dorsally 2 grooved, ventrally 3 grooved, albumen uniform.

Uses : The fruits are used as fish poison.

137. Vitis himalayana (Royle) Brandis Vitaceae

Common Name : Sahmemynain.

Distribution : Throughout N.E. Region.

Large climbers, tendrils, forked, tomentose, Leaves ovate orbicular to triangular - ovate, cordate at base acute to acuminate, serrate, angled or lobed at margins basal nerves 5, lateral nerves 4-7 pairs, cymes or corymbose panicles, peduncles, bearing simple or forked tendrils, Flowers yellow, Berries globose.

Uses : The fruits are dried and the powder is used as fish poison.

138. Xeromphis spinosa (Thunb.) Keay Rubiaceae

(Syn.) *Randia dumetorum* Lamk.

Common Name : Ruthi, Rupanthau, Biskhan hokra.

Distribution : Throughout N.E. Region.

A large deciduous shrub or sometimes a small tree with stout axillary spines. Bark dark - brown or grey, somewhat rough. Leaves generally fascicled, obovate or acute, pubescent, rough or obtusely short, acuminate or acute pubescent, rough or glabrous, Flowers sub-sessile, greenish white or yellowish fragrant, solitary or 2-3 together on short lateral branchlets. Calyx strigose, long campanulate, segments variable ovate or lanceolate or spathulate, foliaceous as long as the tube. Corolla covered with adpressed silky hairs, lobes spreading oval or oblong, stigma fusiform, berry soft, seeds compressed embedded in a gelatinous pulp.

Uses : Fruits are used as fish poison.

139. Zanthoxylum acanthopodium DC. Rutaceae

Common Name : Tambul, Timur, Jaiur, Nangry.

Distribution : Entire N.E. Region in shaded moist localities and foot hills ascending to 1800 m.

A straggling thorny aromatic shrub. Leaves imparipinnate rachis leaflets 15-21 opposite, oblong, lanceolate, caudate - acuminate. Flowers minute greenish white, tetramerous in trichotomous cymes arranged in close set terminal panicles. Fruit globose.

Uses : Fruits are used as fish poison.

7

GUM, RESIN AND TANNIN YIELDING PLANTS

A. GUM PLANTS

140. Acacia farnesiana (L.) Willd. Fl. Br. Mimosaceae

Common Name : Babul, Arimaedah, Benwal.

Distribution : Throughout N.E. Region.

A thorny shrub, with dark - brown smooth bark, branches marked with grey dots and armed with stipular spines. Leaf - rachis long, pinnae 4 - 8 pair. Flowers bright - yellow, exceedingly sweet scented, in globose heads, calyx and corolla 5 toothed.

Uses : Gum is exuded from the bark of the tree which is a good substitute for gum arabic. It is used in confectionery.

141. Adansonia digitata L. Bombaceae

Common Name : Gorakh imili.

Distribution : Throughout N.E. Region.

Branches spreading, Leaves long - petioled, digitately 5 - 7 partite, leaflets lanceolate, downy beneath. Flowers solitary, axillary, calyx cup shaped. Petals 5, white, adnate to the stamens below. Staminal tube short, cylindrical style long, slender, stigmas spreading. Fruit oblong, woody indehiscent pendulous on long stalk, seed brown, reniform, embedded into soft white pulp.

Uses : Pulp of the fruit, bark and leaves are source of gum which is used in medicine.

142. Albizzia lebbek Benth Mimosaceae

Common Name : Amaltas, Chimchola, Sins, Vaaka.

Distribution : All over N.E. Region.

Large spreading trees upto 25 m. high, bark grey, usually cracked, young parts white pubescent, pinnae 6-10 pairs, leaflets 5-15 pairs, oblique-oblong, orbicular, usually with midrib excentric towards upper half, a large gland present near the base of rachis on elevated medium sized near the base of last pair of pinnae, umbel few flowered, peduncle long, slender - staminal tube slightly shorter than corolla tube. Pod linear oblong obtuse, thickened at sutures.

Uses : A reddish brown gum is exuded from the bark. It has smooth polished surface and is used for adulterating gum arabic.

143. Azadirachta indica A - Juss. Meliaceae

Common Name : Neem, Vembu, Vapu.

Distribution : All over N.E. Region.

Shady trees, upto 25 m. tall, bark grey cracked and rough on mature branches. Leaflets 12 - 16 crenate - serrate, oblique upto 6 x 2 cm. Flowers 2-merous, often upto 8- merous. Staminal column 10 - fid at tip, anthers exerted, ovary globose 3 locular, style elongate, stigma 3 lobed with a basal rim. Drupe globose pulpy with a very large seed.

Uses : The bark exudes a clean bright amber coloured gum. It contains a bitter alkaloid "Margosine" in long white needles.

144. Butea monosperma (Lamk.) Taub. Fabaceae

(Syn.) *B. frondosa* Roxb.

Common Name : Pala, Dhak, Bol - aphal - uiz.

Distribution : Throughout N.E. Region upto a height of 300 m.

A deciduous tree with silky young parts, leaves pinnate, trifoliate, leaflets silky tomentose beneath lateral obliquely ovate, terminal rhomboid or broadly ovate, base cuneate, racemes axillary. Flowers bright orange.

Uses : Gum contains recino - tannic and gallic acids.

145. Cyamopsis tetragonoloba Linn. Fabaceae

Common Name : Guar.

Distribution : Throughout N.E. Region as a garden vegetable.

A robust erect annual, 1 - 3 m. high, bearing clusters of thick fleshy pods.

Uses : Source of guar gum which is used in food, paper and textile industry.

146. Delonix regia Rafin Caesalpiniaceae

Common Name : Gulmohar

Distribution : Throughout N.E. Region. Grown for its beautiful flowers in gardens and roadsides.

It is a large nearly evergreen open branched spreading tree, irregularly shaped, more umbrella like with light green foliage. Leaves bipinnate, pinn ie 11-18 pairs. Flowers large, orange/ scarlet.

Uses : The seeds contain a gum which may find uses in textile and food industry.

147. Gardenia turgida Roxb. Rubiaceae

Common Name : Thanella, Duduki, Kutolmaram, Kharhar, Pendro.

Distribution : Throughout N.E. Region.

A small deciduous tree armed with strong axillary often leaf bearing spines, bark smooth, bluish grey, branches mealy. Leaves opposite, crowded at the ends of the branches, very variable, obovate or elliptic, obtuse or short, acuminate, coriaceous, tomentose beneath. Flower axillary or at the ends of short lateral branchlets, corolla white or slightly dilated upwards, lobes oblong, longer than the tube.

Uses : The tree yields a yellow gum.

148. Grevilla robusta A. Cunn. Proteaceae

Common Name :

Distribution : Throughout N.E. Region.

A large tree, young shoots ferruginous tomentose. Leaves alternate pinnate, pinnae 4 - 10, long deeply pinnatifid or entire, the segments usually lanceolate margins recurved, nearly glabrous above silky beneath. Flowers orange coloured in short racemes appearing on short leafless branchlets in the old wood.

Uses : The bark of the tree yields a yellow gum. The gum yields on hydrolysis galactose and arbinose.

149. Lagerstroemia parviflora Roxb. Lythraceae

Common Name : Dhaura, Bondar, Merle, Siddem.

Distribution : Throughout N.E. Region.

A large to medium sized deciduous tree, bark, light brown or ash coloured, smooth, exfoliating in dark coloured, scurfy pieces. Leaves oblong, acute or acuminate, coriaceous, glabrous above, glaucous and very closely reticulate beneath. Flowers white, fragrant in lax axillary and terminal panicles.

Uses : Yields a gum.

150. Mitragyna parvifolia (Roxb.) Korth Rubiaceae

Common Name : Kaim, Kalmi, Mundi.

Distribution : Throughout N.E. Region.

A large deciduous tree, trunk irregularly shaped, often buttressed, bark brownish or greenish grey, smooth with shallow depressions left by exfoliating scales. Flowers light yellow, bracteolate, glabrous inside, in globose heads. Corolla trumpet shaped, lobes 5.

Uses : The plant contains a crystalline alkaloid, gum as well as resin and wax.

151. Sterculia urens Roxb. Sterculiaceae

Common Name : Gular, Gulu, Kalu, Hutim, Kudai, Karay.

Distribution : Throughout N.E. Region.

A moderate sized deciduous tree. Bark greyish white or reddish, smooth peeling of in papery flakes, Leaves crowded at the ends of the branches, palmately 5 - lobed. Flowers greenish yellow, small in terminal panicles, follicles 4 - 6, ovoid oblong.

Uses : Source of gum which contains mucic acid. A clear white gum exudes from the cuts made in the bark throughout the year.

152. Terminalia bellirica Roxb. Combretaceae

Common Name : Bahera, Bhairah, Thani.

Distribution : Throughout N.E. Region.

A large deciduous tree sometimes buttressed at the base. Leaves alternate clustered at the ends of the branches obovate elliptic. Flowers in axillary and extra - axillary drooping spikes from the new shoots.

Uses : Gum is exuded which is demulcent and purgative.

B. RESIN PLANTS

153. Altingia excelsa Noronha Altingiaceae

Common Name : Julili.

Distribution : Assam and adjoining states.

A magnificient tree common in N.E. Region. A large deciduous tree. The wood is reddish - brown with a smooth feel.

Uses : Resin known as storax is obtained from the tree. Storax is a mixture of Cinnamic acid, Cinnamic aldehyde, benzaldehyde, vanilion styrol, styracin etc.

154. Balliospermum montanum Muell. Arg. Euphorbiaceae

Common Name : Jangli Jamalghota, Banbhatti, Danti.

Distribution : Throughout N.E. Region.

A stout undershrub, young shoots pubescent. Leaves alternate, very variable in shape and size, upper lanceolate, lower ovate, often palmately 3 - 5 lobed, firmly coriaceous, acute or acuminate. Flowers small monoecious greenish in short interrupted racemes from the axils of upper leaves. Male flowers 4 - 5 partite membranous, segment orbicular, concave, petals 5.

Female flowers : Sepals 5 ovate pubescent not accrescent, petals 0, disk cupular, thin, ovary 3 celled.

Uses : The resinous oil from the seeds is powerful purgative.

155. Calotropis gigentia (L.) R. Br. Asclepidaceae

Common Name : Madar, Akanda, Erukku.

Distribution : Occurs in waste land throughout N.E. Region.

A large shrub. Leaves simple, opposite, sessile obovate, cordate at the base. Flowers umbels, simple or compound, white.

Uses : A black resin called Madar fluavil, similar to gutta parcha-black acid resin is obtained. The latex to a limited extent is used in tanning industry for decolouring and imparting yellow colour to hide. The latex contains two isomeric resins. A calotropeal and B - calotropeal with acetic and isovaleric acid and B - amyrin.

156. Cannabis sativa Linn. Cannabinaceae

Common Name : Bhang, Siddhi, Ganja.

Distribution : This is very common throughout N.E. Region.

A strong smelling annual of variable height. Leaves stipulate, simple palmatisect, Flowers unisexual, dioecious. Flowers greenish.

Uses : Cannabinol is obtained which is powerful sedative. Hemp oil is used in paint and varnish.

157. Garcinia mangostana Linn. Guttiferae

Common Name : Mongkop, Minbu, Young - zalai.

Distribution : Cultivated in some parts of N.E. Region.

A small or medium sized tree with deep green leathery leaves, Flowers terminal, solitary or in pairs with pink, fleshy petals, Fruits borne near the tips of short branches, mostly on the outer side of the tree.

Uses : Rind contains a bitter resin "Mango stin". It also contains tannin which is removed by boiling the rind.

158. Ipomoea eriocarpa R. Br. Convolvulaceae

Common Name : Kumpan - vel.

Distribution : Throughout N.E. Region.

Wiry trailing herb. Leaves lanceolate triangular or ovate, acuminate or acute or cordate, petiole long. Flowers rosy long and across in sessile or almost sessile capitate cymes. Calyx segments not enlarged in fruit. Corolla red. Stigma rounded, scarcely lobed. Capsule globose hairy, 2 celled, 4 seeded.

Uses : Tuberous root contains a resin similar to Jalap resin. Seeds also contain resin.

159. Justicia adhatoda L./*Adhatoda zeylanica* Acanthaceae

Common Name : Adathodalam, Adusa, Bama, Basinga.

Distribution : Throughout N.E. Region.

An evergreen gregarious shrub 1 - 3 m. high fetid smell. Leaves elliptic or lanceolate, acuminate, entire or undulate or subcoriaceous, Flowers white in dense bracteate peduncled spikes, axillary, solitary or terminal together at the end of the branches, bracts conspicuous, ovate or obovate, stamens 2, hairy ovules 2 in each cell.

Uses : A resin called vasicine is obtained which is used in pharmaceutical preparations.

160. Leonotis nepetaefolia R. Br. Lamiaceae

Common Name : Hejurchej, Dipmal, Dhompo, Merla, Siddem.

Distribution : Throughout N.E. Region.

An ornamental herb or shrub 1.2 - 1.8 m. high, cultivated or found naturalized almost throughout the hotter parts of India. Stem stout puberulous, ovate, crenate - serrate membranous, Flowers in verticillasters, orange - scarlet, bristly, nutlets oblong - obovoid.

Uses : The leaves yield resinic acid upto 7%.

161. Michelia doltsopa Buch. Ham. Magnoliaceae

Common Name : Bara - champ, Safed champ.

Distribution : All over N.E. Region.

A lofty deciduous tree, young parts densely ferruginous tomentose, Leaves ovate, elliptic or oblong lanceolate, silky puberlous underneath. Flowers white.

Uses : Bark contains resin as well as essential oil and tannins.

162. Mirabilis jalapa Linn. Nyctaginaceae

Common Name : Gul-abban, Andhi malligai, Panthrachi, Rangbano.

Distribution : Common in shaded waste places near drains and streams in the valley throughout N.E. Region.

A large erect herb with soft stem and swollen node. Leaves opposite, cordate. Flowers in a 5 lobed calyx like involucre. The colour grades from white, yellow to red.

Uses : Source of resin "Trigonellin".

163. Pinus merkusii Jungh & De Vries Pinaceae

Common Name : Chil, Chilai.

Distribution : Meghalaya, Manipur, Assam.

A tall gregarious tree, branches whorled. Bark thick, deeply and reticulately cracked coming off in thick small plates. Needles long, slender, back convex, sheath persistent, grey, lacerated at apex. Cones ovoid solitary or in two's or three's pyramidal with a blunt knob at the end, wings four times the length of the seed.

Uses : Plant on incision yields an oleo resin.

164. Plumeria rubra Linn. Apocynaceae

Common Name : Gunach, Mir - kong - long - arong, Gulaichi.

Distribution : Throughout N.E. Region.

A small soft wooded tree, bark corky fissured, Leaves oblanceolate fleshy with a intramarginal nerve. Flowers white, fragrant with yellow centre, purplish outsides in terminal peduncled cymes, bracts deciduous, calyx 5 fid. corolla salver shaped, deciduous. Anthers obtuse. Carpels 2, ovate many serrate, stigma lobed.

Uses : It contains resin guercitin which is used in pectoral syrups.

165. Podophylum hexandrum Royle Berberidaceae

Common Name : Bakrachimaka, bhavanbakra, Pupra, Tenumoo-kusoo.

Distribution : Throughout N.E. Region upto a height of 3000 - 4200 m.

An erect succulent herb 35 - 60 cm. high with creeping perennial rhizome, bearing numerous roots, Leaves 2 - 3, orbicular, reniform, peltate with lobed segments, solitary, white or pink cup shaped.

Uses : Source of resin called Podophyllin which is used in medicinal preparations.

C. TANNIN PLANTS

166. Acacia farnesiana Willd. Mimosaceae

Common Name : Babul, Arimaedah, Benwal, Brangdaru.

Distribution : Throughout N.E. India.

A thorny shrub, with dark - brown smooth bark, branches marked with grey dots and armed with stipular spines. Leaf - rachis long, pinnae 4 - 8 pair. Flowers bright - yellow, exceedingly sweet scented, in globose heads. Calyx and corolla 5 toothed.

Uses : Gum is exuded from the bark of the tree which is a good substitute for gum arabic. It is used in confectionery. Bark is a good source of tannin.

167. Acacia torta Linn. Mimosaceae

Common Name : Chilar.

Distribution : Throughout N.E. Region.

A large prickly climbing herb. Young shoots, petioles and inflorescence densely pubescent. Older stem prominently 5 angled. Leaf rachis long with conical pubescent gland near the base. Leaflets 6 - 42 pairs, panicles large. Terminal flowers pale creamy white in globose, pedunculate heads. Pods 6 - 10 seeded.

Flowering : Rainy season, Fruiting : Dec. - Feb.

Uses : Bark contains large quantity of tannin which is used in medicine as well as for tanning.

168. Albizzia lebbek Benth Mimosaceae

Common Name : Khok, Dieng - sah - in, Chinchola, Siris, Vaaka.

Distribution : Throughout N.E. Region, as road side plantation.

A large deciduous tree, leaves bipinnate, rachis with a large cup shaped gland towards the basal half, pinnae usually one pair, rarely two pairs, terminal pair largest, oblong lanceolate, acuminate, base obliquely cuneate to round. Flowers in small peduncled heads.

Uses : The bark contains 7 - 11% tannin and saponin. It is used for tanning fishing nets.

169. Alnus nepalensis D. Don. Betulaceae

Common Name : Intsungtong, Rupupareng, Nep-Udis.

Distribution : Throughout N.E. Region, upto a height of 3000m.

A perennial shrub with rhizomes which are deep orange

brown and are aromatic and pungent. Leaves elliptic or elliptic - lanceolate. Male spike in drooping panicles. Female spikes peduncled, slender in terminal drooping panicles. The fruits are constricted in middle and contain 3 - 6 seeds.

Uses : Bark is a source of tannin. It is used in dyeing to deepen the colour.

170. Bischofia javanica Blume Euphorbiaceae

Common Name : Phangput - araung.

Distribution : Found all over N.E. Region, particularly scattered in valley.

A large deciduous tree with spreading crown. Leaves alternate, 3-5 foliate. Leaflets elliptic or ovate-oblong acuminate, Flowers greenish, dioecious, apetalous in paniculate racemes.

Uses : The twigs and bark contain about 16 % tannin.

171. Careya arborea Roxb. Myrtaceae

Common Name : Petrasing, Kumblai.

Distribution : Assam and Meghalaya.

A fairly large deciduous tree, bark dark grey. Leaves alternate, obovate, oblanceolate or shortly acuminate. Flowers sessile usually a few together in terminal panicles. Stamens numerous in several series, ovary tetralocular rarely pentalocular.

Uses : Bark provides tannins which are also used as medicine.

172. Cassia obtusifolia Caesalpinnaceae

Common Name : Panwad, Phalli.

Distribution : Throughout N.E. Region.

Foetid tall herb. Leaves long, rachis grooved, a linear cylindrical gland present between lowest pair of leaflet. Flowers yellow or white, in sub-sessile pair. Sepals ovate acute spreading. Petals sub-equal, obtuse, the upper truncate. Anthers rounded at the apex. Pod sub-terete, oblique septate, 30-35 seed areole, always exceeding hilum, testa distinctly veined.

Uses : Leaves and Bark are important source of tannin.

173. Casuarina equisetifolia Linn. Casuarinaceae

Common Name : Belatvan = Ben, Jhau.

Distribution : Throughout N.E. Region.

Erect handsome leafless tree upto 30 m tall branches drooping, branchlets green sulcate, drooping in clusters. Male flowers in cylindric spikes at the end of branchlets perianth 0, Stamen 1, Female flowers in small cone like clusters borne at the base of branchlets perianth 0, ovary 1 loculed.

Uses : Bark and leaves are source of tannin and contain 6 - 8 % tannin (Catechol). It is used for tanning hides as it penetrates the hide quickly.

174. Crataeva adansonii Capparidaceae

Common Name : Tikoshak, Tektaak, Barna.

Distribution : Common in Assam and other adjoining states.

Erect or crooked trees upto 20 m tall, branching high above the ground. Petiole with a distinct knob, consisting of glands, leaflets 2-4 time as long as broad, top gradually acuminate with an acute tip, midrib reddish tinged, nerves prominent beneath. Inflorescence terminal on leafy twigs, pedicels leaving their prominent scars on it. Petals clawed. Stamens on androphore and gynoecium on gynophore. Stigma distinct, sessile, berry covered with yellow greyish crust, peeling off later, deep orange seeds dorsally crusted.

Uses : Bark and leaves are source of tannin.

175. Diospyros peregrina (Gaertn) Gurke Ebenaceae

Common Name : Dundo, Gard - phanas, Tarka - kaya.

Distribution : Throughout N.E. Region.

It is a middle sized tree. Leaves alternate, biparous, elliptic or lanceolate to oblong, acuminate. Male flowers clustered whereas female flowers solitary.

Uses : Unripe fruits are source of tannin.

176. Dodonaea viscosa Linn. Sapindaceae

Common Name : Aliar, Santha, Vellari.

Distribution : Throughout N.E. Region.

A stiff evergreen resinous shrub. Flowers polygamous, inconspicuous, yellowish green. Stamens generally 8, ovary 3 - 4 celled. Fruit a compressed capsule.

Uses : Leaves contain 2 acidic resins, gum, albumen, tannin alkloid and saponin, out of the two resins one is insoluble in ether, both however are soluble in chloroform, alcohol and liquid ammonia.

177. Emblica officinalis Gaertn Euphorbiaceae

Common Name : Amluki, Merai, Nelli, Isurkava.

Distribution : Throughout N.E. Region.

A small deciduous tree with smooth greenish grey exfoliating bark. Leaves feathery small narrowly oblong with pinnately arranged leaflets, Fruits depressed globose 1/2 - 1 cm in diameter fleshy and obscurely 6 lobed containing 6 trigonous seeds.

Uses : Bark and leaves are a source of tannin which is used in medicine, Action is astringent.

178. Garcinia mangostana Linn. Guttiferae

Common Name : Mangustan .

Distribution : Throughout N.E. Region.

A small or medium sized tree, 20 - 45 ft. high, with deep green leathery leaves, flowers terminal, solitary or in pairs, with pink fleshy petals, Fruits borne near the tips of short branches, mostly on the outside of the tree, globose, slightly flattened at both ends.

Uses : Rind contains a bitter "Mangostein" resin and tannin. It is obtained by boiling the rind in water and tannin is removed by boiling in alcohol and evaporating. The resulting product is mangostein and resin. Resin is precipitated by redissolving it in alcohol and water.

179. Grevillea robusta A. Cunn. Proteaceae

Common Name : Savukumaram.

Distribution : Throughout N.E. Region.

Tree upto 25 m. high. Leaves fern like, 2 - 3 pinnate, pinnae entire or deeply pinnatifid, lanceolate, dark green above grey silky or white silky beneath. Flowers greenish yellow in showy racemes. Seeds 1 or 2 winged all around.

Uses : The bark of the plant is a useful source of tannin.

180. Hymenodictyon excelsum Wall. Rubiaceae

Common Name : Bhaulon, Bhorkud, Mach.

Distribution : Throughout N.E. Region.

A large deciduous tree. Leaves opposite, elliptic or obovate - oblong, acuminate, often obtuse or rounded, membranous, pubescent, Flower, white fragrant in large terminal panicles with bracteate, spicate branches, pedicels very short, bracts large persistent, Calyx tube short, lobes 5 - 6, deciduous, Corolla, funnel shaped, pilose outside, lobes 5, short valvate, Stamens 5, filaments dilated upwards anther linear. Ovary 2 celled ovules numerous style long exserted, entire stigma spindle - shaped. Fruit a loculicidally dehiscent capsule.

Uses : Contains hymenodictine and bitter substance aesculin.

181. Jatropha curcas Linn. Euphorbiaceae

Common Name : Jangli - erandi, Bon - bheranda, Bag - bharanda.

Distribution : Found in waste places throughout N.E. Region.

A shrub or small tree with a short irregular trunk, young short glandular - tomentose, with an opaque saponaceous juice, bark grey, smooth, peeling off in thin papery flakes. Leaves broadly cordate, usually palmately 5 lobed. Flowers yellowish, in terminal or sub - terminal cymes. Sepals 5, imbricate slightly puberulous, often coloured. Stamens 10, in two series, filaments of the inner series connate, ovary 3 celled, styles 3, glabrous.

Uses : Contains tannin - Jatrophic acid.

182. Lagerstroemia indica Linn. Lythraceae

Common Name : Jerul, Ajhar, Bondar, Merle, Siddem.

Distribution : Throughout N.E. Region.

A shrub with showy flowers. Leaves glabrous, oblong, petals bright - pink, purple or milky white long clawed, much curled.

Uses : Root, bark and leaves are source of tanin. It gives a good fawn colour to light and soft leather.

183. Lannea coromandelica (Houtt.) Merill. Anacardiaceae

Common Name : Jhingan, Mohin, Keked, Madad, Shimli.

Distribution : Throughout N.E. Region.

A moderate sized to large deciduous tree with spreading crown and stout branches, attaining a height upto 24 m. Bark grey or whitish, smooth, exfoliating in irregular rounded plates. Leaves imparipinnate 25 - 45 cm long, crowded at the ends of branches, leaflets 7 - 9, oblong ovate, Flowers small yellowish or purplish, unisexual male fascicled in slender compound racemes. Female in simple pubescent racemes fruit red, compressed.

Uses : The bark contains 8% tannin. It can be used both for tanning and dyeing purposes. The bark extract dyes cotton with or without mordant and imparts shade.

184. Mallotus philippinensis Muell Arg. Euphorbiaceae

Common Name : Kamala, Rottigra, Kamalaguri.

Distribution : Throughout N.E. Region a common tree found in ravines or in open country by the side of the drains.

A small evergreen tree. Leaves alternate ovate oblong lanceolate, acuminate pubescent and with numerous close set orbicular red glands beneath. Flowers small dioecious, Male flower in terminal spikes, female solitary.

Uses : Source of important resin called rottlerin and tannic acid.

185. Nerium indicum Mill. Apocynaceae

Common Name : Kaner, Karker, Rajbaka. Karab - baha.

Distribution : Throughout N.E. Region.

A branching evergreen shrub. Leaves ternate, linear lanceolate, tapering at both ends, coriaceous. Flowers in terminal cymes, rose or white coloured.

Uses : Tubers contain "Neriodorin" and "Neriodorein" and tannic acid which are used as medicine.

186. Nyctanthes arbor-tristis Linn. Oleaceae

Common Name : Har, Harsinghar, Kharkhasa,Manjappu.

Distribution : Throughout N.E. Region.

Small deciduous tree or bushy shrub, branchlets rough, pubescent 4 angular, bark rough brown, greyish or greenish. Leaves ovate shortly acuminate, distantly toothed or entire coriaceous, scabrid lateral nerves 4 - 6 on either half. Flowers several, scented, sessile, 3 - 7 together on hairy quadrangular peduncles of various length which are arranged in cymes or cymose panicles, bracts elliptical, Fruit capsule.

Uses : Leaves contain an alkaloidal principal named Nyctanthine and tannins.

187. Pithecellobium dulce Benth. Mimosaceae

Common Name : Arda, Morogmara, Thorearang, Mengel, Pani.

Distribution : Assam, Meghalaya, Manipur.

A small middle sized tree, leaves bipinnate, rachis angled with a large gland near the base, pinnae 2 - 4 pairs, with a gland just below the insertion of each pair of leaflets. Leaflets lanceolate to elliptic or rhomboid ovate.

Uses : The bark contains tannin of catcheol type the contents of which vary according to age of the plant. The bark can be advantageously used as a substitute for wattle bark. The bark also contains a yellow dye and 1.5 % pectin.

188. Polygonum plebeium R. Br. Polygonaceae

Common Name : Beng = Chemti sag, dubia sag, Mui - ara.

Distribution : Throughout N.E. Region.

Prostrate polymorphic herbs, root stock often woody. Leaves linear oblong or rarely obovate sessile, apiculate. Stipule hyaline, lacerate and fimbriate upto middle, pedicel jointed just before perianth. Perianth, outer 2 subacute suffused with green outside, inner 3 obtuse dark red. Stamens 3 or fewer.

Uses : Root is source of Polygonic acid, Tannic acid and Gallic acid. The whole plant yields about 11 % tannin.

189. Spondias pinnata (L.f.) Kurz Anacardiaceae

Common Name : Amara, Jangli, Amna, Amoa, Dieng - soh - per.

Distribution : Throughout N.E. Region.

Straggling or erect trees, branchlets stout. Leaves alternate, glabrous leaflets 7-13 oblong, acuminate entire, with an intramarginal nerve. Flowers 5 nerved greenish - white polygamous, clustered on stout branches of panicles. Disc 10 lobed with 8 - 10 stamens, drupe ovoid or oblong, fleshy.

Uses : Leaves and bark are a source of tannin. Gum is also exuded from bark with smooth shining surface.

190. Syzygium cuminii Linn. Skeels Myrtaceae

Common Name : Jamun, Jam, Jambul.

Distribution : Imphal valley and throughout N.E. States.

A large tree, leaves variable elliptic, oblong or ovate, petiole channelled. Flowers sessile, greenish white in panicles of cymes.

Uses : Bark contains 12 % tannin and also tannic, gallic and Tellogic acid and gum.

191. Tamarindus indica Linn. Caesalpiniaceae

Common Name : Imli, Ambli, Beng = Tentul.

Distribution : Assam and other adjoining states.

A large tree. Leaves abruptly pinnate, stipules small, caducous. Leaflets 10 - 20 pair. Flowers in few flowered, subterminal racemes. Calyx tube turbinate, teeth lanceolate, the two lowest connate. Petals pale - white, variegated with red stripe, the middle one hooded, the two lower reduced to scales, stamens only 3 perfect, monadelphous, the others reduced to bristles, anthers versatile. Ovary stipitate, the stalk adnate to the calyx tube. Ovules many, stigma capitate.

Uses : Bark contains tannins. Pulp of the fruit contains tartaric, citric and malic acid.

192. Terminalia arjuna (Roxb.) Wt. & Arn Combretaceae

Common Name : Arjun, Kohu, Pere, Kava.

Distribution : Throughout N.E. Region.

Medium to large sized tree upto 40 m tall. Leaves oblong or elliptic hard, coriaceous, glabrescent spikes solitary, axillary and in terminal panicles. Flowers pale yellow.

Uses : Bark contains tannin named Glucotanic acid upto 15 %. The fruits contain 7 - 20 % tannin.

193. Woodfordia fruticosa Kurz. Lythraceae

Common Name : Dheira, Dhai, Dhavai, Icha - Baha, Juji.

Distribution : Throughout N.E. Region.

Large spreading shrub with reddish brown bark, peeling off in thin strips, blaze pink, young parts hairy, leaves and branchlets covered with black dots. Leaves sessile or sub - sessile, opposite sometimes in whorls of three distinct nerves, lanceolate or ovate - lanceolate, acuminate scarlet, shortly pedicellate, zygomorphic, in short panicles of closely clustered cymes.

Uses : Bright red flowers contain 20% tannins.

8

TIMBER PLANTS

194. Acacia catechu (L.f.) Willd.

Mimosoideae

Uses : Source of catachu obtained from heart wood, a product called 'katha'.

195. Acacia ferruginea (L.) Willd.

Mimosoideae

Uses : Wood is used for cart wheel, post beams and agricultural implements, tree also yields gum.

196. Acacia leucophloea (Roxb.) Willd.

Mimosoideae

Uses : Bark used for tanning, wood used for agricultural implements.

197. Acacia sundra DC. Khair

Mimosoideae

Uses : Wood used for beams, posts and agricultural implements. Tree yields a gum with considerable tracts in kutch. It is sold in Bombay under the name "Red abong".

198. Adenia sessifolia

Passifloraceae

Uses : Juice of the leaves and root is used in skin trouble. Root used as an ingredient of tonic.

199. Adina cordifolia (Roxb.) HK. f.

Rubiaceae

Uses : The wood is used for tea chests and other minor uses.

200. Aegle marmelos (L.) Corr.

Mimosaceae

Uses : Unripe fruits are astringent and are used in dysentery. Pulp is aromatic and cooling used in the form of 'sherbat'.

201. Albizzia lebbek Benth

Mimosoideae

Uses : Wood is excellent for high class furniture, interior decoration.

202. Ailanthus excelsa Roxb.

Simaroubeceae

Uses : Wood is used for catamaran and for small boat.

203. Alstonia scholaris (L.) R. Br.

Apocynaceae

Uses : Furniture, Plywood, Writing boards.

204. Amoora wallichi King.

Meliaceae

Uses : Wood used for boat building and railway carriages and also suitable for furniture, decorative work and plywood.

205. Amoora rohituka Wight & Arn.

Meliaceae

Uses : Wood used for boat building and for furniture.

206. Anogeissus acuminata (DC.) Guill. & Perr.

Bakla, Dhavada, Vellanora.

Combretaceae

Uses : Wood strong and elastic, as good as hickory and ash wood for tool handles.

207. Anogeissus latifolia (DC.) Wall ex Guill & Perr.

Combretaceae

Uses : Tree yields a gum which is a good substitute for gum arabic used in pharmaceutical preparations.

208. Anthocephalus cadamba (Roxb.) Miq

Rubiaceae

Uses : Wood used chiefly for packing purposes and tea boxes, also employed for beams and rafters and light construction.

209. Aquilaria agallocha Roxb.

Thymelaeaceae

Uses : Source of agar which consists of irregular patches of dark wood highly charged with oleoresin found in old trees.

210. Azadirachta indica A. Juss.

Meliaceae

Uses : Bark used in skin troubles. Leaves considered as antiseptic. Decoction given for ulcer and eczema.

211. Barringtonia - acutangula (L.), Gaertn.

Lacythidaceae

Uses : Tender leaves edible. Bark, root and seeds employed as fish poison. They probably contain saponin. Powdered seeds are used as an expectorant. Bark contains tannin.

212. Bauhinia malabarica Roxb.

Caesalpinaceae

Uses : Young shoots and leaves are edible.

213. Bauhinia purpurea Roxb.

Caesalpinaceae

Uses : Roots carminative. Bark used in diarrhoea, also yields a fibre. Flower buds eaten as a pot herb.

214. Bauhinia racemosa L.

Caesalpinaceae

Uses : Bark yields a strong cordage fibre. Bark is astringent and used in dysentery. Leaves given with onions for diarrhoea as an anthelmintic.

215. Bauhinia retusa Roxb.

Caesapinaceae

Uses : Tree yields gum known as "sapila" and used for sizing cloth, paper and as binder for charcoal.

216. Bischifolia javanica Bl.

Euphorbiaceae.

Uses : Wood used for construction purposes like house posts. being especially durable under water. Also used for cheap pencils.

217. Boswellia serrata Roxb. ex. Colebr.

Burseraceae

Uses : Wood used for making news print and other types of paper. Also used for plywood, venner, hard boards, cheap furniture.

218. Buchanania lanzen Spreng.

Anacardiaceae

Uses : Seeds are eaten which are nutritious. The seeds yield a fatty oil and substituted for olive and almond oil.

219. Butea monosperma (Lamk.) Taub.

Fabaceae

Uses : Young shoots yield a fibre used for ropes. Much used for cabinet work, brush backs, agricultural implements, poles and carts.

220. Callicarpa arborea Roxb.

Verbenaceae

Uses : Wood used for charcoal making.

221. Camellia sinensis (L.) O. Kuntze.

Saatong, Singo

Theaceae

Uses : Cured leaves are used for preparation of tea.

222. Canarium bengalense Roxb.

Burseraceae

Uses : Source of resin used as incense Wood does not wrap and can be used for planks, shingles and tea boxes.

223. Casearia graveolens Dalz.

Samydaceae

Uses : Wood suitable for carving. Fruit used as fish poison.

224. Casearia tomentosa Roxb.

Samydaceae

Uses : Bark used in tanning. Fruit is diuretic, powdered bark applied in dropsy.

225. Cassia fistula L.

Caesalpinaceae

Uses : Dried fruit used as purgative, laxative for habitual constipation.

226. Cassia simea L.

Caesalpinaceae

Uses : Wood used for cart wheels, Posts, beam and agricultural implements. Tree yields a gum.

227. Chloroxylon sweitenia

Rutaceaee

Uses : Wood durable, not liable to insect and fungus attack, used for cabinet work, brush backs, agricultural implements, poles, carts and rail road.

228. Cinnamomum cecidaphne Meissn.

Lauraceae

Uses : Wood strongly scented, highly prized for cabinet. Also used for construction work, planks, rafters, ploughs and yokes. Tree yields an essential oil.

229. Cleistanthus collinus (Roxb.) Benth. ex HK. f & Th.

Euphorbiaceae

Uses : Wood used for house posts. Leaves aborifacient. Root, leaves and bark used as fish poison.

230. Cochlospermum tomentosum Kunth

Cochlospermaceae

Uses : Root used as fabrifuge and tonic, juice of plant used by the Arabs in the preparation of a fermented beverage.

231. Cordia myxa Roxb.

Boraginaceae

Uses : Fruit is astringent, anthelmintic diuretic, demulent and expectorant, used in disease of chest and urinary passage.

232. Cynometra polyandra Roxb.

Caesalpiniaceae.

Uses : Wood suitable for heavy constructional work, hammers and tool handles, paving blocks and boat building.

233. Dalbergia latifolia Roxb.

Fabaceae

Uses : Wood used for tool handles and agricultural implements. Decoction or bark used in dyspepsia, seed oil is used for rheumatism. Bark contains tannin.

234. Dichrostachys cinerea Wight et Arn.

Fabaceae

Uses : Wood used for walking sticks.

235. Dillenia pentagyna Roxb.

Dillenaceae

Uses : Wood used for home posts, rafters and planking.

236. Diospyros chloroxylon Roxb.

Ebenaceae

Uses : A good fodder plant. Fruit edible.

237. Diospyros embropteris Pers.

Ebenaceae

Uses : Ripe fruits edible. Raw fruits contain tanin.

238. Diospyros melanoxylon Roxb.

Ebenaceae

Uses : Source of coromandel ebony which consists of the black heart wood, substituted for true ebony. Leaves diuretic and laxative.

239. Dipterocarpus macrocarpus Vesque.

Dipterocarpaceae

Uses : Used in the manufacture of plywood or tea chests also used as railway sleepers.

240. Dipterocarpus turbinatus Gaertn. f.

Dipterocarpaceae

Uses : Wood used for internal constructional work, packing cases. tea boxes, panelling and flooring, also for carriages and wagons. Source of kanyin oil of Burma and gurjun oil of Bengal.

241. Dolichandrone falcata Seem.

Bignoniaceae

Uses : Bark yields a course fibre, used also as a fish poison. Wood used for building and agricultural implements.

242. Elaeodendron glacum Jacq. f.

Celastraceae

Uses : Wood useful for cabinet works, combs, picture frames. Powdered leaves employed as snuff, yields a gum Jamrasi. Bark and leaves contain 8.0 - 13.5% tannin.

243. Emblica officinalis Gaertn.

Euphorbiaceae

Uses : Fruit sour and astringent, cooling diuretic, laxative eaten raw or cooked, also pickled. A rich source of vitamin C. containing twenty times as much vitamin 'C' as orange juice.

244. Erythrina indica Lamk.

Fabaceae

Uses : Leaves and tender shoots consumed as a pot - herb, leaves also used as cattle fodder. Leaves considered laxative diuretic, anthelmintic applied externally in venereal buboes.

245. Erythrina suberosa Roxb.

Fabaceae

Uses : Wood used for scabbards, laddles, jars for storing ghee, drum, lacquered boxes, water troughs and packing cases. Also considered suitable for match boxes and yields fibre suitable for cordage.

246. Eucalyptes hemipholia f. Muell.

Myrtaceae

Uses : Excellent timber used for rail road sleepers, bridges, poles, plankings, handles, wheels.

247. Eugenia jambolana Lam. (Syn). *Syzygium cumini* (L.) Skeels

Myrtaceae

Uses : Ripe fruit edible. Spirituous liqour as well as a wine is prepared from them. Fruit also used for making preserves, jams, squashes and jellies.

248. Feronia elephantum Correa.

Rutaceae

Uses : Fruit edible. Also used for making 'sharbat' considered tonic, antiscorbiatic and alexipharmic. Leaves aromatic, carminative, astringent. Also yields essential oil. Tree lopped for fodder, word used for agricultural implements.

249. Ficus religiosa L.

Moraceae

Uses : Fruits and tender buds eaten in time of scarcity. Hard latex is used to fill up cavities in hollow ornaments. Bark contains tannin. Wood used for packing cases.

250. Flacourtia - ramontchi L.

Flacourtiaceae

Uses : Fruit edible, appetizing and digestive used in Jaundice and enlarged spleen. Bark astringent and diuretic. Wood used for agricultural implements and posts.

251. Gmelina arborea Roxb.

Verbenaceae

Uses : Wood used for furniture, planking, Venetian blinds, tight cooperage carriages, printing blocks, carving musical instruments, shaft axels, lacquered, boxes, picture frames, artificial limbs.

252. Gyrocarpus jacquini Jacq.

Hernandiaceae

Uses : Wood used for furniture, planking, Venetian blinds, tight cooperage carriages, printing blocks, carving musical instrument shafts, axels, lacquered, boxes, picture frames, artificial limbs.

253. Gyrocarpus amoricanus Jacq.

Hernandiaceae

Uses : Wood used for toys, school models, combs, trays, boxes, and furnitures, specially suitable for catamarans and paddles.

254. Hardwickia pinnata Roxb. Anjan.

Caesalpiniaceae

Uses : Wood used for beams, rafters. Tree yields an oleo resin which is a source of volatile oil used as substitute for imported clove oil.

255. Holarrhena antidysenterica (L.) Wall.

Apocynaceae

Uses : Bark is astringent, anthelmintic stomachic.

256. Holoptelea integrifolia Roxb.

Ulmaceae

Uses : Wood used for brush back and handles of dusting brooms. Also used for indoor building purposes.

257. Hymenodictyon excelsum Roxb. Wall.

Rubiaceae

Uses : Fruit edible and also used for making 'sharbat'. Fruit is considered as tonic. Tree lopped for fodder. Wood used for tea boxes.

258. Ixora arborea Roxb. ex. Sm.

Rubiaceae

Uses : Wood is used for furniture and building purposes.

259. Kydia calycina Roxb.

Malvaceae

Uses : Wood used for planks, agricultural implements. Also used for ply wood.

260. Lagestroemia parviflora Roxb.

Lauraceaee

Uses : Wood used for building construction, fencing, bridges and decorative furniture.

261. Lannea grandis Engl.

Anacardiaceae

Uses : Wood used for home buildings, packing cases and furniture.

262. Litsea polyantha Juss.

Lauraceae

Uses : Wood used for furnitures, tea boxes and ply wood.

263. Machilus gamblei, (Syn). *Presea gamboloi* (King ex HK. f.) Kosterm.

Lauraceae

Uses : Wood suitable for house building, bridges and furniture. Bark used as bitter tonic in fever.

264. Macaranga peltata (Roxb.) Muell - Arg.

Euphorbiaceae

Uses : Yields a reddish gum 'kino' used for sizing papers and for taking impressions of coin.

265. Madhuca latifolia (Koen.) Macbr.

Sapotaceae

Uses : Seeds are source of mahua oil used chiefly in the manufacture of laundry soap and to some extent for edible purpose after refining. Also used in rheumatism and skin diseases.

266. Mallotus philippinensis (Lamk.) Muell. - Arg

Euphorbiaceae

Uses : Fruit yields a dye called kamla powder which gives a rich golden red colour on silk. Also employed as anti oxidant for ghee. Bark contains tannin.

267. Mangifera sylvatica Roxb.

Anacardiaceae

Uses : Unripe fruit suitable for tarts, pickles and jellies, wood used for same purpose as mango wood.

268. Manihot utilissima Crantz

Euphorbiaceae

Uses : Fruit yields a dye called 'kamla' powder giving a rich golden red colour to silk. Also employed as an oxidant for ghee. Bark contains tannin.

269. Michelia champaca L.

Magnoliaceae

Uses : Wood used for posts, furniture, carriage and ship building.

270. Millingtonia - hortensis Linn. f.

Bignoniaceae

Uses : Wood suitable for furniture, tea boxes and drawing board etc.

271. Mimusops elengi L. Bakaula

Sapotaceae

Uses : Wood used for building purposes, bridges, boats etc.

272. Mitragyna - parviflora (Roxb.)

Rubiaceae

Uses : Wood used for agricultural implements, furniture, carving and match boxes.

273. Morinda tinctoria Roxb.

Rubiaceae

Uses : Wood used for furniture, toys, slate frames, yokes etc.

274. Neohouzeaua dollooa (Gamble) Camm.

Gramineae

Uses : Weed used for umbrellas, baskets and mats.

275. Nyctanthes arbor-tristis Linn.

Oleaceae

Uses : Flowers are most important and are used as offering and for extracting essential oil.

276. Odina woodier Roxb.

Anacardiaceae

Uses : Wood is used for house building.

277. Oroxylum indicum Linn.

Bignoniaceae

Uses : Root and bark juice is tonic and astringent. Used in diarrhoea and dysentery and rheumatism. Wood used for match boxes.

278. Oxytenanthera nigrociliata Munro.

Gramineae

Uses : Suitable for paper-pulp.

279. Phoebe goalparensis Hutchins.

Lauraceae

Uses : Principal source of bonsom wood used for house construction, planking, furnitures and cabinet work.

280. Phoenix sylvestris Roxb.

Palmaceae

Uses : Tree trapped for its sap, used for making Jaggery. Sugar sap called nira is a refreshing drink. Root used for toothache.

281. Plectronia dydima (L.) Kurz.

Rubiceae

282. Polyalthia cerasoides (Roxb.) Bedd.

Annonaceae

Uses : Wood used for house construction, planks, rafters, packing cases.

283. Polyalthia longifolia Thw.

Annonaceae

Uses : Wood used for barrel, drums and packing cases for scaffolding and carriages. Bark febrifuge .

284. Pongamia pinnata (L.) Pierre

Fabaceae

Uses : Seeds yield a fatty oil. Pongam oil, used in tannin industry for dressing. It is also used in the preparation of washing soap and candles. Bark yields fibre used for cordage.

285. Premna bengalensis C.B. Clarke

Verbenaceae

Uses : Bark soft, sweet and edible, old wood lasts well in water and used for bridges and house posts also suitable for turnery and carving.

286. Premna tomentosa Willd.

Verbenaceae

Uses : Leaves diuretic and vulnerary used in dropsical infection. Decoction given after parturition. Leaves yield an essential oil.

287. Prosopis spicigera (L.) Druce

Mimosaceae

Uses : Wood suitable for interior decoration and construction work and also agricultural implements.

288. Pterocarpus indicum Willd.

Fabaceae

Uses : Wood used for tea boxes and other light packing cases suitable also for planking and plywood. Bark contains tannin.

289. Pterocarpus marsupium Roxb.

Fabaceae

Uses : Wood used chiefly for building purposes also doors, window and frames.

290. Pterygota alata (Roxb.) R. Br.

Sterculiaceae

Uses : Wood used for tea boxes. Bark yields a fibre used for rough cordage.

291. Randia dumetorum (Thunb.)

Rubiceae

Uses : Fruit edible after roasting or cooking. They are emetic and used as a substitute for Ipecacuanha in small doses, expectorant and diaphoretic. Flowers yield an essential oil.

292. Rhododendron arboreum Sm.

Ericaceae

Uses : Wood used chiefly for tea boxes light packing cases. The other possible uses are in ship building, and match boxes.

293. Saccopetalum tomentosum Bennet

Annonaceae

Uses : Wood used mostly for building huts and cattle sheds and also suitable for cabinet work. Tree yields a gum.

294. Salmalia malabarica (De) Schott & Endl.

Bombacaceae

Uses : Wood is most widely used in match industry, specially for match boxes. Also suitable for shingles and toys. Tree yields a gum called mochares.

295. Santalum album L.

Santalaceae

Uses : Source of Indian sandal wood oil extracted from the heart wood. Sandal wood is one of the finest wood for carving and also employed for curios of exquisite beauty.

296. Sapindus emargianatus Vahl.

Sapandaceae

Uses : Wood used for charcoal making. Fruit contains saponin which is used in textile industry.

297. Saraca indica L.

Caesal piniaceae

Uses : Wood is used for ploughs and shafts.

298. Schima wallichii (DC.) Korth.

Theaceae

Uses : Wood used for building purposes

299. Schleichera oleosa (Lour.) Oken.

Sapindaceae

Uses : Wood used for poles, construction & cabinet works.

300. Schrebera swietenioides Roxb.

Oleaceae

Uses : Wood used for agricultural implements, construction and cabinet works.

301. Semicarpus anacardium Linn. f.

Anacardiaceae

Uses : Fleshy orange cup (hypocarp) of the fruit eaten when ripe, significantly astringent pericarp abounds in a black oily, bitter and highly vesicant juice which has been traditionally used for marking linen.

302. Shorea robusta Gaertn.

Dipterocarpaceae

Uses : Sal wood ranks with teak and doeodar as one of the best sleeper wood in India, also in great demand in form of poles.

303. Soymida febrifuga A. Juss.

Meliaceae

Uses : Bark used in the treatment of diarrhoea, dysentery and fever and also as a general tonic.

304. Spondias mangifera Willd.

Anacardiaceae

Uses : Wood employed for packing cases and tea chests, floats and boats, non ornamental ply wood and fairly good as unbleached wood pulp.

305. Sterculia urens Roxb.

Sterculiaceae

Uses : Wood is used for packing cases.

306. Sterculia villosa Roxb.

Sterculiaceae

Uses : Yields a gum known as gum karaya.

307. Stereospermum chelonoides (L.f.) DC.

Bignoniaceae

Uses : Wood is used for canoes & heavy packing cases.

308. Stereospermum personatum (Hassk) D. Chatterjee

Bignoniaceae

Uses : Wood used for building purpose. bridge, planking and agricultural implements. Bark used for dyeing.

309. Stereospermum suaveloens (Roxb.) DC.

Bignoniaceae

Uses : Wood employed for construction work and furniture.

310. Swietenia mahogoni Jacq.

Meliaceae

Uses : Wood highly prized for decorative furniture.

311. Strychnos potatorum L.

Loganiaceae

Uses : Mostly used for medicinal purposes.

312. Tamarindus indica L.

Caesalpinaceae

Uses : Weeds used as tonic, stomachic, demulcent and emetic.

313. Terminalia arjuna (Roxb. ex DC.) Wt & Arn

Combretaceae

Uses : Wood used for carts, agricultural implements. It may also be used for house building.

314. Terminalia bellirica (Gaertn.) Roxb.

Combretaceae

Uses : Wood used for tough work. Wood is only of second quality.

315. Terminalia chebula Retz.

Combretaceae

Uses : Wood used for tough work. Seeds used as medicine and for tanning. Kernel yields fatty oil.

316. Terminalia tomentosa Wt. & Arn.

Combretaceae

Uses : Bark contains tannin and may be used for tanning. Wood used for beams, rafters, doors and windows.

317. Tetrameles nudiflora R. Br.

Tetramelaceae

Uses : Second quality wood.

318. Trema orientalis Blume.

Ulmaceae

Uses : Wood used for tea chests and match boxes.

319. Trewia nudiflora L.

Euphorbiaceae

Uses : Wood used for purposes where much strength is not required such as tea boxes, light packing cases and match sticks.

320. Vitex peduncularis Wall. ex Schauer

Verbenaceae

Uses : Wood used in posts and beams.

321. Wrightia arborea (Roxb.) Roem. & Schult

Apocynaceae

Uses : A second class wood, used for wood items.

322. Xylia xylocarpa (Roxb.) Taub.

Mimosaceae

Uses : This is one of the most important timber used for sleeper.

323. Xylia dolarbriformis Benth.

Mimosaceae

Uses : Wood, stronger and heavier than *X. xylocarpa.*

324. Zanthoxylum budranga (Roxb.) DC.

Rutaceae

Uses : Wood used for planking. Best for carving.

FOOD AND EDIBLE PLANTS 9

325. Acetephila excelsa (Dolz.) Muell Arg.

Euphorbiaceae

Leaves used as vegetable.

326. Actinidia callosa Lindl.

Actinidicaea

Ripe acidic pulp edible.

327. Aegle marmelos Correa ex Roxb.

Rutaceae

Aromatic pulp edible. Pulp of the fruit is used as pickle and chatni.

328. Aesandra butyracea (Roxb.) Baehni

(Syn.) *Diplo knema butyracea* (Roxb.) Lem.

Sapotaceae

Fat extracted from the seed is edible.

329. Aglaia edulis A. Gray

Meliaceae

Sweet succulent fleshy aril edible. It is regarded as cooling and astringent.

330. Alpinia galanga Willd.

Zingiberaceae

The green rhizome contains .04% essential oil. Leaves also yield a volatile oil. Rhizome used in soup like preparations.

331. Alpinia malaccensis Rosc.

(Syn.) *Catimbium malaccense* (Bum. f.)

Zingiberaceae

The rhizomes yield a volatile oil. The rhizomes are reported to be used in Java for sores. The fruit infused with a little salt are emetic.

332. Alpinia nigra (Gaertn.) Burtt

Zingiberaceae

Rhizomes are used as vegetable.

333. Alpinia speciosa K. Schum.

Zingiberaceae

The fibrous stems contain 49.9% of cellulose and can be used for making paper. The rhizome of the plants used as a substitute for ginger.

334. Amaranthus caudatus Linn.

Amaranthaceae

It is cultivated on hill slopes for the starchy edible seeds which the Monpas use as cereal like rice. Its seeds constitute the food grain "Ram dana".

335. Amorphophallus campanulatus (Roxb.) Bl. ex Decne

Araceae

It is cultivated for their edible corms. Their acrid and irritating taste is due to presence of calcium oxalate.

336. Ampelocissus barbata Wall.

Vitaceae

Sweet berries taste like grapes.

337. Angiopteris evecta (Forst.) Hoffm.

Marathaceae

Underground parts and base of the rachis stem eaten after boiling and thorough washing.

338. Antidesma acuminatum Wall.

Euphorbiaceae

Sub sweetish pulp edible.

339. Antidesma diandrum (Roxb.) Heyne ex. Roth

(Syn.) *Antidesma acidum* Retz.

Euphorbiaceae

Leaves and fruits used as vegetable.

340. Antidesma thwaitesianum Muell. Arg.

Euphorbiaceae

Sub - sweetish pulp edible.

341. Alpinia rubra (Roxb.) Radlk.

Zingiberaceae

Ripe fruits are edible.

342. Aporusa dioica Muell. Arg.

Euphorbiaceae

Acid fruit aril eaten raw.

343. Ardisia crispa

Myrsinaceae

Young shoots and leaves are acidic and are eaten either raw or cooked.

344. Ardisia griffithii Clarke

Myrsinaceae

In Meghalaya flowers are cooked and eaten.

345. Ardisia polycephala Wall.

Myrsinaceae

Young shoots and leaves are acidic in nature and are eaten raw or cooked.

346. Ardisia solanaceae Roxb.

Myrsinaceae

Young shoots and leaves are acidic and are eaten either raw or cooked.

347. Artocarpus heterophyllus Lamk.

Moraceae

Fruits are pickled and are also used as vegetable. The seeds are also eaten after roasting. It is reported that an extract of the heartwood with alum is used by Buddhist priests for dyeing their ropes yellow.

348. Asparagus racemosus Willd.

Liliaceae

Roots are eaten after boiling. This is regarded a delicacy. The raw roots are also used in number of ayurvedic preparations.

349. Asystasia neesiana (Wall.) Nees.

Acanthaceae

Leaves are used as vegetable.

350. Averrhoa carambola Linn.

Oxalidaceae

The fruits are made into stews, curries, pudding and tarts. The sweeter ones are eaten as dessert. The slighty green ones are made into jams, jellies, pickle and preserves. The flowers are sometime made into preserve. Carambola crush can be a refreshing drink.

351. Baccaurea sapida Muell. Arg.

(Syn). *Baccaurea ramifolia* Lour.

Euphorbiaceae

Sweetish pulp edible. The yellow globose fruit about an inch in diameter possesses a pleasant acidic taste.

352. Bambusa bambos (L.) Von.

Gramineae

Young buds are cooked as vegetable.

353. Bambusa tulda Roxb.

Gramineae

Young shoots are pickled and eaten. *B. tulda* from the USA was found to yield 46% of unbleached pulp.

354. Bauhinia purpurea L.

Caesalpiniaceae

Flower - buds are cooked and used as vegetable.

355. Benincasa hispida (Thunb.)

Cucurbitaceae

The fruit is used as vegetable.

356. Bombax ceiba L.

Bombacaceae

Flowers, buds, fleshy calyx eaten raw.

357. Buddleja asiatica Lour.

Loganiaceae

Leaves used in the preparation of a drink called Laupani.

358. Bupleurum falcatum Linn.

Umbelliferae

Boiled roots are used as vegetable by Khasis

359. Calamus rotang Linn.

Arecaceae

Young shoots and culm are pickled. The young tender shoots of some species are used as vegetable. The root is used in Cambodia as a remedy for dysentery and biliousness and febrifuge and tonic. It is also used in veterinary practice as aperient.

360. Capsicum annum Linn.

Solanaceae

Fruits are used as vegetable and also as spice. Capsicum preparations are used as counter irritants in lumbago, neuralgia and rheumatic disorder.

361. Casearia esculenta Roxb.

(Syn).*Casearia bourdillonu* Mukherjee

Somydaceae

Leaves and young shoots are edible. The leaves are used in stews, and in medicated bath.

362. Castanopsis tribuloides (Smith) A. DC.

Fagaceae

Seeds are edible.

363. Celtis australis Linn.

Ulmaceae

Sweetish pulp eaten. The leaves are used as fodder. The fruit is sweet and edible. The seeds yield a fatty oil.

364. Chenopodium album Linn.

Chenopodeaceae

The plant contains and ethereal oil, a substance resembling cholesterol and ammonia and amine both in free and combined form. It is used as vegetable.

365. Chlorophytum arundinaceum Baker

Liliaceae

Leaves are used as vegetable. The root is used as a tonic in Eastern Himalayas and Assam.

366. Cirsium lipskyi Petrak

Asteraceae

Leaves and young shoots eaten, by the Khasis, Mikirs and also by Manipur tribals. Its aromatic seeds and young shoots are edible.

367. Citrus hystrix DC.

Rutaceae

Juicy acidic fruits eaten raw.

368. Citrus medica Linn.

Rutaceae

The fruits are used mainly for pickling. The juice is refrigerant and astringent.

369. Clausena dentata (Willd.) Roem & Schult

Rutaceae

Sweet fruit pulp tastes like black current. The leaves are aromatic. The essential oil obtained from the leaves is of a yellow colour and has an agreeably weak fruity odour.

370. Clerodendrum colebrookianum Walp

Verbenaceae

Leaves and tender shoots edible.

371. Clerodendrum serratum (L.) Moore

Verbenaceae

Leaves used as vegetable.

372. Codonopsis ovata Benth

Campanulaceae

Leaves and young shoots are eaten cooked by Khasis.

373. Colocasia esculenta (Linn.) Schott

Araceae

Corm is edible.

374. Conocephalus suaveolens Bl.

Moraceae

Leaves and tender shoots eaten cooked.

375. Coptis teeta Wall.

Ranunculaceae

The rhizome is an all round tonic. It has been found efficacious in debility.

376. Cordia wallichii G. Don.

Borginaceae

The fruit is considered to be expectorant, astringent and demulcent.

377. Corlaria nepalensis Wall.

Coriariaceae

Insipid fruit pulp is eaten.

378. Cornus capitata Wall. ex Roxb.

Cornaceae

Pulp sweet, eaten raw and also made into preserves.

379. Costus specious (Koening) Sm.

Zingiberaceae

The rhizome is edible and is used after cooking. It is mucilaginous and feebly astringent.

380. Crataegus oxycantha Linn.

Rosaceae

Acidic pulp eaten raw or as preserve.

381. Cucumis sativus Linn.

Cucurbitaceae

The seeds are reported to be cooling tonic and diuretic. The seeds are edible and are used in confectionery. They contain protein and fat.

382. Cucurbita maxima Duch.

Cucurbitaceae

All parts of the plant are edible. Tender shoots and leaves are eaten in salads. The flowers are cooked and eaten as vegetable.

383. Curcuma zedoaria Rosc.

Zingiberaceae

The rhizomes are rich in starch. The "choti starch" of commerce is a product extracted from the rhizomes and used as a substitute for arrowroot and barley.

384. Cyanotis tuberosa Roem & Schult

(Syn). *C. concanensis* Hamk.

Commelinaceae

Leaves and tender shoots are edible.

385. Cythocalyx martabanicus HK. f. Th

Annonaceae

Aromatic sweetish pulp eaten

386. Dendrocalamus hamiltoni Nees & Arn.

Gramineae

Shoots boiled and eaten. The inner layer clum - sheath is used for wrapping cigarettes. In Burma young shoots are used as vegetable.

387. Dichopsis polyantha Benth. & HK. f.

(Syn). *Palaquium polyanthum* (Walt) Baill.

Orchidaceae

Boiled flowers are eaten.

388. Dillenia indica Linn.

Dilleniaceae

Sweetish pulp edible.

389. Dillenia pentagyna Roxb.

Dilleniaceae

Flower buds are cooked with dry fish.

390. Diospyros kaki Linn. f.

Ebenaceae

Sweetish pulp edible. Roasted seeds can be used as a substitute of coffee.

391. Dioscorea bulbifera Linn.

Discoreaceae

Tubers are eaten after removing the bitter and acrid substance by boiling and thorough washing in running water. The flowers are used as vegetable.

392. Dioscorea glabra Roxb.

Discoreaceae

The tubers are consumed in Andaman Island and Khasi hills but are not much liked as they become gluey when cooked.

393. Dioscorea puber Blume

Discoreaceae

The tubers are edible and are reported to be good though some varieties emit an offensive odour when cooked.

394. Dioscorea pentaphylla Linn

Discoreaceae

Tubers are edible after removing the bitter and acrid substance by boiling and thorough washing in running water. The flowers are often collected and used as vegetable.

395. Dioscorea trinervis Roxb. ex. Prain & Bark.

Discoreaceae

Tubers eaten boiled.

396. Dioscorea wallichii HK. f.

Discoreaceae

Tubers eaten after boiling and thorough washing.

397. Docynia hookeriana Decne

Rosaceae

Fruits sub - sweet, consumed raw and also as preserve.

398. Donella roxburghii (G. Don) Pierre ex. Lecomte

Sapotaceae

Sweet pulp edible.

399. Duchesnea indica (Andr.) Focke

Rosaceae

Sour pulp edible, fruit juicy.

400. Ehretia acuminata R. Br.

Boraginaceae

The ripe fruit is insipid and edible, the unripe fruit is pickled. The leaves are used as fodder. The tender leaves of the closely related var. *obovata* (Lindl.) Johnston are reported to be used for mixing with tea.

401. Elaeocarpus varunua Buch. Ham.

Elaeocarpaceae

Fruit is eaten both raw and cooked.

402. Elaeagnus latifolia Linn.

(Syn). *E. injindibularis* Momiyamu

Elaeogonaceae

Sub - sweetish pulp eaten, also used in tarts and jellies.

403. Elatostema dissectum Wedd.

Urticaceae

Leaves and fruits edible both as raw and cooked.

404. Elatostema platyphyllum Wedd.

Urticaceae

Leaves consumed as vegetable.

405. Eleocharis dulcis Trin.

Cyperaceae

Small tubers edible.

406. Eleusine coracana Gaertn.

Gramineae

Seeds edible and also used for making local drink.

407. Embelia gamblei Kurz.

Myrsinaceae

Young leaves and shoots consumed after boiling.

408. Embelia sessifolia Kurz.

Myrsinaceae

Sub - sweetish pulp eaten raw.

409. Emblica officinalis Gaertn.

Euphorbiaceae

Fruits eaten raw and also pickled.

410. Ensete superbum (Roxb.) Cheesman

Flower buds cooked as vegetable.

411. Entada phaseoloides Merrill

Mimosaceae

Large seeds roasted/ boiled and eaten. The seeds are considered tonic, emetic, antiperiodic and anthelmentic. Two amorphorous saponins which are nearly alike in chemical and pharmacological properties have been isolated from the seed.

412. Entada pursaetha DC.

Mimosaceae

The juice of the pounded paste is used for washing hair and claimed to eradicate dandruff.

413. Eryngium foetidum Linn.

Umbelliferae

Plants cooked with vegetables for flavouring.

414. Erioglossum rubiginosum Blume

Sapindaceae

The tender shoots of the plant are used as vegetable in Java. Leaves and roots are used in Malaya for sub - sweetish pulp.

415. Eriosema chinense Vog.

Fabaceae

The tuberous roots which are of the size of marble are eaten raw.

416. Eugenia formosa Wall.

(Syn). *Syzygium formosum* (Wall.) Masamune

Myrtaceae

Calyx cooked as vegetable.

417. Eugenia kurzii Duthie

Myrtaceae

Sub - sweetish pulp eaten.

418. Eulophia campestris Wall.

Orchidaceae

Small tuber eaten raw and also boiled.

419. Euryale ferox Salisb.

Nymphaeaceae

The seed eaten raw or roasted. On roasting in hot sand the seed coat swells and bursts and can be easily peeled off.

420. Evodia froxinifolia (D. Don) HK. f.

Rutaceae

Sub - sweetish pulp edible, used in chutney.

421. Fagopyrum cymosum Meissn.

Polygonaceae

From the seed the Monpas prepare flour and chapati (bread). It is principally a fodder plant. The leaves are cooked and eaten as vegetable.

422. Feronia limonia (Linn.) Swingle

(Syn). *Limonia acidissima* L.

Rutaceae

Aromatic pulp sub - sweetish, eaten raw or made into drinks.

423. Ficus auriculata Lour.

Moraceae

Fruit cooked with curries.

424. Ficus hispida Linn.

Moraceae

The fruit which are pale green or greenish yellow are edible when ripe and are made into jam.

425. Ficus hirta Vahl.

Moraceae

Young leaves and fruits are edible.

426. Fissistigma polyanthum Merrill.

Annonaceae

Aromatic, sub - acidic pulp edible.

427. Flacourtia jangomas (Lour.) Raeusch

Flacourtiaceae

Flowers eaten raw or as preserve. Pulp with tarty.

428. Flemingia vestita Benth.

Fabaceae

The tubers are nearly elliptical and possess an agreeable nutty flavour.

429. Gallingsoga parviflora Cav.

Asteraceae

The plant is reported to be used as a vegetable.

430. Garcinia atrovirdis Griff.

Guttiferae

The fruit rind is too sour to be eaten raw but tastes excellent when stewed with sugar.

431. Garcinia lanceaefolia Roxb.

Guttiferae

The leaves are sub - acid and reported to be eaten by Mikirs after cooking.

432. Garcinia paniculata Roxb.

Guttiferae

The aril of the fruit, like that of mangosteen is highly flavoured and is eaten with relish. The plant has been recommended as a suitable root stock for magosteen.

433. Garcinia xanthochymus HK. f.

Guttiferae

The fruit has a juicy pulp with a pleasant acid flavour. It is made into preserves and jams.

434. Gardenia campanulata Roxb.

Rubiaceae

The leaves and fruit are eaten after cooking.

435. Gardenia latifolia Ait

Rubiaceae

Sub - sweetish pulp edible.

436. Gaultheria fragrantissima Wall

Ericaceae

Sweetish pulp edible.

437. Gelonium multiflorum Juss.

Euphorbiaceae

Sweetish pulp edible.

438. Gironniera cuspidata Bl. Kurz.

Ulmaceae

The drupe is eaten and the foliage is used as fodder.

439. Gnetum gnemon Linn.

Gnetaceae

The seeds are eaten after roasting or cooking. The orange or red pulp of the fruit is removed and the seed kernel mashed and moulded into cake which is dried in the sun and fried in boiling oil into a kind of cake or biscuits.

440. Haematocarpus thompsonii Miers

Menispermaceae

Sweet pulp eaten, fruit very juicy.

441. Hodgsonia heterociliata HK. f. & Th.

Cucurbitaceae

Seeds roasted and kernel eaten, much liked by Mizo and Manipur tribals.

442. Hodgsonia macrocarpa (Bl.) Cogn.

Cucurbitaceae

Seeds washed, roasted and eaten.

443. Horsfieldia amygdalina (Wall.) Warb

Myrsticaceae

Fruit aril sweet, edible.

444. Houttuynia cordata Thunb.

Sauraceae

Small tuber eaten raw, also used in curries

445. Hovenia dulcis Thunb.

(Syn). *H. acerba* Lindl.

Rhamnaceae

Succulent peduncles, inflorscence, taste like pear.

446. Hygrophila salicifolia Nees

Acanthaceae

Leaves and shoots cooked as vegetable.

447. Illicium griffithii HK. f. & Th.

Magnoliaceae

The fruits are used as spice due to their aromatic and carminative properties.

448. Impatiens amplexicaule Edgew

Balsaminaceae

The oil extracted from seed is used as varnish.

449. Indigofera dosua Buch.-Ham. ex. D. Don

Fabaceae

Flowers cooked as vegetable.

450. Ipomoea aquatica Forsk.

Convolvulaceae

Leaves and shoots are used as vegetable

451. Kydia calycina Roxb.

Malvaceae

The lac of this plant is used as dye for colouring clothes.

452. Lasia spinosa L. Thw.

Araceae

Small tuber and leaves used as vegetable.

453. Leea asiatica (L.) Ridsdale

(Syn). *Leea crispa* L.

Vitaceae

Fruit insipid, edible.

454. Leea indica (Burm. f.) Meir.

Vitaceae

Tender leaves and shoots are used in curries.

455. Lepisanthes rubiginosa Roxb.
Sapindaceae
Ripe fruit edible.

456. Lindera neesiana (Nees) Kurz
Lauraceae
The oil extracted from the seed is edible.

457. Litsea cubeba (Lour.) Pers.
Lauraceae
Fruit yields an edible oil.

458. Litsea glutinosa (Lour.) Robins
Lauraceae
Fruit insipid, edible.

459. Litsea kingii HK. f.
Lauraceae
Fruit yields oil.

460. Lysimachia candida Lindl.
Primulaceae
Tender leaves and shoots are used as vegetables in Manipur.

461. Madhuca latifolia(Koen) Macbr.
Sapotaceae
Seeds yield edible oil.

462. Maesa argentea Wall.
Myrsinaceae
Fruit insipid, and that also of *M. chisia*.

463. Maesa indica Wall.
Myrsinaceae
Young leaves/shoots are used as vegetable.

464. Maesa montana DC.
Myrsinaceae
Young leaves/shoots are used as vegetable.

465. Mahonia acanthifolia G. Don.

Berberidaceae

Pulp sour and edible.

466. Malachara capitata Linn.

Malvaceae

Fruits and leaves used as vegetable.

467. Malva verticillata Linn.

Malvaceae

Young leaves/shoots used as vegetables.

468. Mangifera indica L.

Anacardiaceae

Pulp sour, sub - sweetish, much relished eaten raw/pickled also of *M. sylvatica*.

469. Mangifera sylvatica Roxb.

Anacardiaceae

Fruit pickled, also cooked in curries like *M. indica*.

470. Medinilla rubicunda Bl.

Melastomataceae

Cooked leaves edible.

471. Melastoma malabathricum Linn.

Melastomataceae

Fruit insipid, edible tastes like black berry.

472. Meliosma pinnata Roxb.

Sabiaceae

Leaves tender/shoots cooked with fish or vegetable by Mikirs. Fruits edible.

473. Melodinus monogynus Roxb.

Apocynaceae

Berries fleshy, sub - sweetish, edible.

474. Merremia umbellata (Linn.) Hall.

Convolvulaceae

Leaves used as vegetable.

475. Mesua ferrea L.

Clausiaceae

Fruit insipid, edible.

476. Meyna spinosa Roxb.

Rubiaceae

Fruits and young leaves are edible.

477. Monochoria hastaefolia Presl.

Pontederiaceae

Tender leaves used as vegetable.

478. Morinda umbellata Linn.

Rubiaceae

Fruit used in curries.

479. Moringa oleifera Lamk.

Moringaceae

Fruit pickled. Tender pods used as vegetable.

480. Morus alba Linn.

Moraceae

Fruits are edible.

481. Mucuna prurita (L.) DC.

Fabaceae

Fruit (young green pods) cooked as vegetable.

482. Murraya koenigii (L.) Spreng

Rutaceae

Leaves are used as spice.

483. Mussaenda frondosa Linn.

Rubiabeae

Young leaves/sprouts are edible.

484. Mussaenda glabra Vahl.

Rubiaceae

Tender leaves eaten raw and as salad and chutney.

485. Mussaenda roxburghii HK. f.

Rubiaceae

Dye obtained from the leaves. Leaves used as vegetable.

486. Myrica esculenta Buch.- Ham. ex D. Don.

Myricaceae

Fruit eaten raw or made into a drink.

487. Myrsine capitellata Wall.

Myrsinaceae

Fruit insipid, edible.

488. Natsiatun herpeticum Buch. - Ham.

Icacinaceae

Young leaves/shoots used as vegetable.

489. Nelumbo nucifera Gaertn.

Nymphaeaceae

Rhizome used in soups and curries.

490. Nymphoides indicum (L.) Kuntze

Nymphaeaceae

Rhizome used in soups and curries.

491. Oenanthe javanica (Bl.) DC.

Umbelliferae

Young leaves/shoots cooked as vegetables and also used raw.

492. Olax acuminata Wall. ex Benth.

Olacaceae

Young leaves/shoots cooked as vegetable by Mikirs.

493. Olax nana Wall. ex Benth.

Olacaceae

Fruit insipid, edible.

494. Oxalis martiana Zucc.

Oxaldiaceae

Small tuber eaten raw.

495. Parthenocissus himalayna (Royle) Planch.

Vitaceae

Pulp sub - sour edible.

496. Paedaria foetida Linn.

(Syn). *P. scandens* (Lour) Merill

Rubiaceae

Tender leaves boiled and made into stews/curries.

497. Parkia roxburghii G. Don.

(Syn). *P. timorina* (A. DC) Merill

Mimosaceae

Fruit (young green pods) cooked as vegetable.

498. Pavetta subcapitata HK. f.

Rubiaceae

Leaves cooked as vegetable.

499. Pegia nitida Colebr.

Anacardiaceae

Pulp sub - sour, aromatic, edible.

500. Pennisetum americanum (L.) Leeke

Gramineae

Extensively cultivated for edible seeds which are boiled, or sometimes fried.

501. Persea kurzii (King ex HK. f.) Kosterm

Lauraceae

Valuable timber tree. Plant juice is used as a drink.

502. Phaseolus torosus Roxb.

Fabaceae

Used as vegetable, commonly cultivated on hill slopes.

503. Phoenix acaulis Buch. - Ham. ex Roxb.

Palmaceae

Fleshy - sweet pulp edible.

504. Phytolacca acinosa Roxb.

Phytolaccaceae

Leaves/shoots cooked as vegetable, also cultivated in higher hill by Arunachal tribals.

505. Picris hieracioides L.

Asteraceae

Leaves/shoots cooked as vegetable.

506. Pinanga hookeriana Bece

Palmae

Fleshy sweet pulp edible.

507. Polygonatum multiflorum (L.) Ait.

Liliaceae

Starchy tubers cooked as vegetable.

508. Polygonatum verticillatum (L.) Ait.

Liliaceae

Starchy tubers are used in curries and soup by Manipur tribals.

509. Polygonum runcinatum Buch.-Ham.

Polygonaceae

Cooked flowers and inflorescence eaten.

510. Pouzolzia viminea Wedd.

Urticaceae

Leaves edible and are used by Khasis as vegetable.

511. Premna latifolia Roxb.

Verbenaceae

Leaves/tender shoots edible.

512. Protium serratum (Wall. ex Colebr.) Engl.

Burseraceae

Sub - sweetish pulp edible.

513. Prunus cerasoides D. Don.

Rosaceae

Sub - sour/acidic fruit edible.

514. Prunus cornuta (Royle) Steud.

Rosaceae

Sub - sour acidic fruit edible.

515. Pteridium aquilinium (L.) Willd.

Polypodiaceae

Young sprouting frond cooked as vegetable.

516. Pterygota alata (Roxb.) R. Br.

Sterculiaceae

Roasted seeds are edible.

517. Pyrus pashia Buch. - Ham. ex D. Don

Rosaceae

Pulp edible and tastes like pear.

518. Pyrularia edulis (Wall.) A. DC.

Santalaceae

Fruit tastes like guava, edible.

519. Randia spinosa (Thunb.) Poir.

Rubiaceae

Berries edible occasionally eaten by Assam tribals.

520. Randia uliginosa (Retz.) DC.

Rubiaceae

Fruit and young green pods cooked as vegetable.

521. Rhus javanica L.

(Syn). *R. semialata* Murray

Anacardiaceae

Sub - acidic pulp much relished.

522. Rhynochotechum ellipticum A. DC.

Gesneriaceae

Leaves cooked as vegetable.

523. Ribes glaciale Wall.

(Syn). *R. takare* D Don.

Saxifragaceae

Fruits are edible.

524. Rorippa indica (L.) Heirn.

Cruciferae

Used as vegetable and oil is also obtained.

525. Rosa gigantea Collett.

Rosaceae

Soup made out of flowers.

526. Rourea minor (Gaertn.) Alston

Connaraceae

Red aril edible.

527. Rubus biflorus Buch. - Ham. ex Sm.

Rosaceae

Sweet juicy berries edible.

528. Rubus rugosus Sm.

Rosaceae

Fruit juice used for curing fistula, fruits edible.

529. Rumex maritimus Linn.

Polygonaceae

Leaves/tender shoots cooked as vegetable and soup like preparations.

530. Sagittaria sagittifolia Linn.

(Syn). *Sagittaria trifolia* Linn.

Alismaceae

Tubers eaten after boiling.

531. Salacia roxburghii Wall.

(Syn). *Salacia salacioides* (Roxb.) Rao & Hemadri

Hippocrateaceae

Pulpy gelatinous mass edible.

532. Sambucus javanica Reinw. ex Bl.

(Syn). *Sambucus hookeri* Rehder

Caprifoliaceae

Tender leaves cooked as vegetable.

533. Sarcochlamys pulcherrima (Roxb.) Gaud.

Urticaceae

Tender leaves/young shoots cooked as vegetable.

534. Saurauia cerea Griff. ex Dyer.

Dilleniaceae

Sub - sour berries edible.

535. Smilacina fusca Wall.

Liliaceae

Leaves used as vegetable.

536. Solanum barbisetum Nees.

Solanaceae

Pulpy Berries edible.

537. Solanum crassipetalum Wall.

Solanaceae

Tender leaves used as vegetable.

538. Solanum kurzii Brace ex Prain

Solanaceae

Fruit (young green pods) used as vegetable.

539. Solanum melongena L.

Solanaceae

Used as vegetable.

540. Solena heterophylla Lour.

(Syn). *Solena amplexicaulis* (Lamk.) Gandhi

Rubiaceae

Leaves used as vegetable

541. Sorbus aucuparia Linn.

Rosaceae

Pear like fruit, (sub - sweetish) edible.

542. Sterculia guttata Roxb.

Sterculiaceae

Seed edible after roasting.

543. Sterculia indica Merrill

Sterculiaceae

Fruit (young green pods) used as vegetable.

544. Stixis suaveolens (Roxb.) Pierre

Capparidiaceae

Sweet aromatic pulp edible.

545. Streblus indicum (Bur.) Corner

Moraceae

Insipid fruit edible.

546. Syzygium aqueum Alston

Myrtaceae

Sub - acidic pulp edible.

547. Tetrastigma lanceolarium (Roxb.) Planch

Vitaceae

Sub - acidic berries edible.

548. Trichosanthes dioica Roxb.

Cucurbitaceae

Fruit (young green pods) used as vegetable.

549. Trigonostemon semperforeus

Euphorbiaceae

Fruit edible.

550. Typha angustifolia Linn.

Typhaceae

Young shoots boiled or stewed and used in various ways.

551. Uraria cordata (Dunal) Alston

Fabaceae

Fruit insipid edible

552. Vaccinium donianum Wt.

Vacciniaceae

Young leaves used as vegetable.

553. Vaccinium griffithianum Wt.

Vacciniaceae

Sub-acidic pulp edible.

554. Vaccinium serratum Wt.

(Syn). *Vaccinium vacciniaceum* (Roxb.) Sleumer.

Vacciniaceae

Flowers sour, used in curries.

555. **Vaccinium sprengeli** Wall.
Vacciniaceae
Sbu - acdic pulp edible.

556. **Viburnum colebrookianum** Wall. ex Clarke
Caprifoliaceae
Sub - sour fruit edible.

557. **Viburnum cotinifolium** Don
Caprifoliaceae
Sub - sour fruit edible.

558. **Viburnum grandiflorum** Wall.
Caprifoliaceae
Sub - sour fruit edible.

559. **Vigna pilosa** Baker
Fabaceae
Seeds edible after boiling - used as pulse.

560. **Vigna sinensis** (L.) Savi ex Hassk.
Fabaceae
Commonly cultivated for seeds. Used as vegetable and pulse.

561. **Vigna umbellata** (Thunb.) Ohwi & Ohashi
Fabaceae
Seeds edible after boiling. Used as pulse.

562. **Vigna vexillata** A. Rich
Fabaceae
Tuber of small carrot size, eaten raw and boiled.

563. **Vitis lanata** Roxb.
Vitaceac
Succulent berries edible, sub - sweetish.

564. **Willughbeia edulis** Roxb.
Apocynaceae

Acidic pulp edible.

565. Ximenia americana Linn.

Olacaceae

Sub - sweetish pulp edible.

566. Zanthoxylum oxyphyllum Edgew.

Rutaceae

Tender shoots used as vegetable.

567. Zingiber zerumbet (L.) Rosc. ex Smith

Zingiberaceae

Rhizomes give good flavour and are used in soup.

568. Zizyphus apetala HK. f.

Rhamnaceae

Sub - sweetish pulp edible.

569. Zizyphus funiculosa Ham.

Rhamnaceae

Sub - sweetish pulp edible.

570. Zizyphus incurva Roxb.

Rhamnaceae

Sub - sweetish pulp edible.

REFERENCES

Ambasta, S.P. (Ed.) 1986. The useful plants of India C.S.I.R., New Delhi

Anonymous 1948 - 1976. The wealth of India - Raw materials, Vol. 1 - 11 C.S.I.R., New Delhi.

Anonymous 1976. Botanical Exploration Tawang R.R.L., Jorhat, Assam, India.

Arora, R.K. Native Food Plant of North East Tribals. Glimpses of Indian Ethnobotany, Ed. S.K. Jain, Oxford and I.B.H. Publishing Co. New Delhi. pp. 91-106.

Bennet, S.S.R. 1983. Ethnobotanical Studies in Sikkim, Indian For. 109 : 477 - 481 (12).

Bennet, S.S.R. 1985. Ethnobotanical Studies in West Sikkim, J. Econ. Tax. Bot. 7 : 317 - 321 (52).

Deb, D.B. 1961. Diocotyledonous Plants of Manipur Territory Bull. Bot. Survey. India, No. 3 and 4, 253 - 350.

Deb, D.B. 1960. Medicinal Plants of Tripura State. Indian For. 94 : 753 - 765.

Dam, D.P. and Hajra, P.K. Observations on Ethnobotany of the Monpas of Kameng District, Arunachal Pradesh. In : Jain S.K. (Ed.) 149 - 152. Glimpses of Indian Ethnobotany. Oxford and I.B.H. Publishing Co. New Delhi. pp. 107-114.

Hajra, P.K. 1981. Nature Conservation in Khasi folk beliefs and taboos. ibid. pp. 149-152. ibid.

Hajra, P.K. and Baishya, A.K. Ethnobotanical Notes on the Miris (Mishings) of Assam plains. pp. 161 - 170 ibid.

Joseph, J. and Kharkongor, P. A Preliminary Ethnobotanical Survey in the Khasi and Jaintia Hills, Meghalaya. pp. 115 - 123 ibid.

Kanjilal, U.N. et al. 1936 - 1940. Flora of Assam, Vol. IV. Calcutta.

Krishna. B. and Singh, S. 1987. Ethnobotanical Observation in Sikkim, J. Econ. Tax. Bot. 9 : 1 - 7 (18).

Kumar, S. 1986. Non - Conventional Legumes of North Eastern States (in press).

Kumar, S. 1995. Medicinal Plants of North Eastern Region (in press).

Kharkongor, P. and Joseph, J. Folk Lore Medico - Botany of Rurāl Khasi and Jaintia Tribes in Meghalaya. pp. 124 - 136.

Glimpses of Indian Ethnobotany, Ed. S.K. Jain, Oxford and I.B.H. Publishing Co. New Delhi.

Rao, R.R. Ethnobotanical Studies on the Flora of Meghalaya, Some Interesting Reports of Herbal Medicines. pp. 137 - 148 ibid.

Rao, R.R. 1993. Floristic diversity of Eastern Himalaya - a national heritage for conservation Sym. Vol. V. Dhar Himalayan Biodiversity Conservation Strategies, G.B. Plant Himalayan Institute, Almora.

Singh, P.K., Singh, N.I. & Singh, L. J. 1988. Ethnobotanical studies on wild edible plants in the markets of Manipur. I.J. Econ. Tax. Bot. 6 : 699 - 703. (16).

Vasudeva, M.K. and Shanpru, R. Some Plant in the Life of the Guros of Meghalaya. Glimpses of Indian Ethnobotany, Ed. S.K. Jain, Oxford and I.B.H. Publishing Co. New Delhi.

Walls, G. 1889 - 99 A Dictionary of Economic Products of India, Vols 1 - 6, London.

INDEX OF BOTANICAL NAMES